AF327312

About the Author

Jim Goedert is retired after teaching 22 years at the University of Nebraska at Kearney. At the present time, Jim serves on the Board of Directors for both the Kansas Barbed Wire Museum at LaCrosse, Kansas and the Devil's Rope Museum in McClean, Texas. He is also the founder and co-chairman of the four-year old "Barbed Wire Symposium." Jim was elected to the Barbed Wire Hall of Fame in 1998. He is also the director of the Kearney Softball League.

About the Author

Larry Greer, a native Kansan, is a retired inventor and designer. As research historian for the barbed wire hobby since 1970, he has written over 400 barbed wire related articles and has written four books and contributed to nine others. He was inducted into the Barbed Wire Hall of Fame in 1982.

PLANTER WIRE

A Patent History and Collector's Catalog

BY JIM GOEDERT AND LARRY GREER

Contents

Dedication

We dedicate this book to the American farmer, who not only fed himself, but included the rest of us at his dinner table, the farmer-inventors who made this possible, and the planter wire collectors who are true "Stewards of the Past." They search out lost and forgotten relics and preserve them so the rest of us and future generations can not only see, but can touch and understand a bit of our history.

Larry Greer and Jim Goedert
February, 1998

Acknowledgements

Jim and Larry wish to acknowledge those who have offered their help in making this book possible: Harold Hagemeier, Amarillo, Texas, for his sketches; Brad Penka, curator of the Kansas Barbed Wire Museum, for allowing access and opportunity to photograph the planter wire collections; Gerald Huebert, Portsmouth, Iowa, for his first person accounts; Rick Tworek, Case Corporation for the Case G-32 manual; Dick Cutler, Forest Lake, Illinois, for the Deere and Mansur 999K manual: Marvin Ginn, Morrowville, Kansas for allowing his planter wire collection to be photographed; Ron Larsen, Kearney, Nebraska for the quality photography; Billie Thornton, Aurora, Colorado for his endless search of materials; Mary Elliott Roessler and Dee Urwiller Zeller for their many hours of proofreading and typing; and Angel Bauer for the actual lay out of the book.

Special thanks to Dee Urwiller Zeller, University of Nebraska at Kearney, for guiding us to the elusive first Check-line patent; Larry's daughter, Tanya for copy assistance; and Larry's wife, Coleen, without whose editing skills and encouragement, this book might not have been possible.

Introduction

The main purpose of the 1996 Barbed Wire and Tool Symposium held at the Kansas Barbed Wire Museum in LaCrosse, Kansas, was how best to advance the hobbies of barbed wire, planter wire, fencing tools and all related items. During the meetings, a request was made for new reference books for collectors. Chairing the session, Delbert Trew, curator of the Devil's Rope Museum, McLean, Texas, asked for volunteers to take on various projects. Jim Goedert and I agreed to produce a planter wire book. I would do the research and write the text. Jim would seek out specimens from collections, obtain photographs, oversee the production of the book, and underwrite the project.

While still in LaCrosse for the symposium, I started searching for planter wire patents at the Kansas Barbed Wire Museum using the *U.S. Patent Indices*. I discovered that very few patents exist exclusively for planter wire, more commonly referred to as check-line or check-row wire. I did find over 1,000 patents titled "check-row planter" and others simply titled "corn planter." Returning home, I searched through my own patent files and those in the St. Louis University Law Library and added several hundred more patents to the list. From this list I selected 750 patents that seemed most likely to yield clues to a check-line or knot.

When the patent copies arrived, the real task began. It soon became clear that in the development of most planters, the check-line was a lesser factor in the patent. It was sometimes shown in the figures and described in the text but seldom specified in the claims. There are just a few patents exclusively for the check-line or knot. Even though check-line planters relied entirely on the check-line to trip the drop mechanism of the seed box, most patentees that came after the first few seemed to take the check-line for granted.

Some planter patents show unique forms of knots and check-lines buried within the figures. Others show indistinguishable check-lines, but described within the text is enough detail about the lines and knots to present a good idea of their forms. From this information sketches were created that most likely match the spirit of the patent. The specimen figures in this book are drawn from three sources: (1) the patent object figures, (2) sketches either enhancing vague figures or interpreted from the text, and (3) sketches of specimens in collections. For easier identification most illustrations approximate the actual size of the knot or check-line as closely as it is possible to ascertain.

This book contains illustrations of nearly 200 planter wire examples. Many of the patents suggest that there are other variations possible. Variations from the patent figures are often the result of the manufacturing process, economics or on-going improvements to the product. Some variations are the result of other individuals attempting to copy or modify a patented design to suit their own planter designs. For these reasons we will never be able to identify all planter wire specimens with an exact patent match. Therefore, an unidentified specimen may be named according to its shape, or the location where it was found, or the name of the collector who found it, or a variation of a known patent until new evidence suggests otherwise.

The earliest mention of check-row planting found in the patent reports occurred in the 1840s. The era of the check-line check-row planter began in 1857 when a farmer, Martin Robbins, from Hamilton, Ohio, was granted patent 16,611, the first known check-line planter patent. The last patent evidence found of the use of check-lines was patent 2,175,035, issued in 1939 to C.K. Shedd for a four-row planter. The patent was assigned to the Iowa State College Research Foundation. Between those dates the entire check-line check-row planter industry was born, flourished and almost disappeared. However, the International Harvester Company was still making the McCormick No. 240 Two-Row Check-Row Planter as late as 1952. There were many manufacturers of check-row planters that used check-lines. Most of these were located in Iowa, Illinois, Indiana, Kentucky, Wisconsin and Ohio, but the greatest number of inventors lived in Illinois where corn was gaining status as the major crop.

Corn was not the only crop that was planted using the check-row method. The check-line has been used in planting cotton and potatoes as well. The row distances differed from corn rows, but the procedure was the same. The check-row method could be used on most crops that required evenly spaced plants and sufficient room around them for cultivation.

It was over ten years after Robbins' patent before the use of check-lines came into general acceptance. The Civil War undoubtedly was a factor in this delay. By 1870, G.D. Haworth, recognizing the value of the early efforts of Robbins and others, adapted the check-line principle to his corn planter. It was not long before other early inventors made check-row attachments to upgrade their planters. In this book the Haworth name appears frequently. The Haworth family of Decatur, Illinois acquired the patent rights by assigned re-issue from some of the early inventors. Those patents, along with several of their own, gave them dominance in the new industry. The Haworths were involved with various planter patents from 1861 through 1901.

The knot on a section of check-line is the object that collectors seek. Few of the old corn planters exist and are not conducive to collecting except by museums. However, the knot is an excellent representative of the planters that used it, and it is certainly more easily found and collected. Dozens of different examples exist, and from these collectors are creating interesting and historic collections.

The knot on the check-line was an obstruction that caused a trip-lever to actuate the seed drop mechanism as the planter was pulled along. In some cases a spoked, rimless wheel was employed instead of a lever. In that instance, the knot engaged the end of the spoke, causing the wheel to rotate that in turn actuated the seed drop. In both designs, the planter slipped past the knot, and the lever or spoke was readied for the next knot.

Some patents specified more and some specified fewer knots per row spacing. That was because of competing designs for check-row attachments, and the search for the optimum knot design that would operate the trip levers or wheels without causing mechanical disruptions. For example, there were check-lines with two knots closely spaced, one to trip the drop lever, the other to reset it. The knot spacing or added function was the claimed feature of the patent. A few patents used the same knot found in other patents. The claim in those patents was for the use of more knots, but not the knot itself. Many knots that are now found appear similar because of aging, so, in order to distinguish one patent from another, it may be necessary for the collector to polish or dissect the knot to determine its true construction.

Some knots served as line connectors, others as splices for broken lines. Knots acted as throw-offs to drop or disengage the check-line from the planter when it was necessary to turn around at the edge of the field. There are knots so large or so shaped that one can imagine the problems that occurred as they went through the pulleys and seed dropper mechanisms. Other knots were only intended to act as visual indicators for hand actuating the seed dropper or for hand planting.

There are also different types of check-lines: single and multiple wire, cord, rope, chain and strip. Some check-lines are continuous, while others are chain-like. A single knot on the check-line, centered in an 18-inch length, is the standard specimen length. This is a carry-over standard from the barbed wire hobby. At the 1996 Symposium it was agreed to also allow a shorter specimen of 4 1/2 inches, because it was believed that the single knot specimen was sufficient proof of a patent feature. However, research shows that some patents will require more than 18 inches to show the patent feature. This presents an extension to the hobby for those who choose to further expand their collections to include multiple knot specimens representing the patents that claim multiple or fractional knot spacing. With this in mind, a collector should exercise caution before dividing a find. There are now three acceptable lengths: 4 1/2 inches for a shorty specimen, 18 inches for a standard specimen and sometimes over 18 inches for a two knot specimen, depending on the patent.

This book includes photographs and sketches of specimens in collections that illustrate many of the variations and unknowns that have been found. One section covers some of the reels and check-line anchors used. Another section is miscellaneous planting implements dated from 1850 to 1939. The section on check-row planter manuals and advertisements will provide additional understanding of the check-row process. Three indices have been provided to make it easier to locate the different subjects. One is the chronological order of check-line and knot patent numbers. A second index is of the inventors of check-lines and knots in alphabetical order. The third index is for the check-line anchors, reels and related items.

Do not become discouraged if you do not find an exact match for your specimen. Should no identification come close, it is acceptable to declare the specimen "unknown." It will still add interest to your collection and some mystery as well.

While researching, considerable effort was made to locate the patent identifications for all known knots, however, they will never all be conclusively resolved. Because research within the hobby is continuous, and more identities are sure to surface, feel free to contact Jim Goedert or me for assistance in identification concerning new finds. We are also interested in any patent information, check-row planter manuals, old planter advertisements, etc. for possible future publication. Happy hunting.

Larry J. Greer
February, 1998

Definitions

The following definitions concern check-line check-row planters and their related U.S. patents.

Anchor --A stake or other means used to secure the end of the check-line.

Ball -- A wood or metal object, spherical or oblong, used as the enlargement in a check-line.

Button -- A perforated metal disk spaced on the check-line to suit the distance between the rows being planted. It is the enlargement in a check-line.

Chain -- A wire check-line usually made of wire links equal to the distance between corn rows and joined by interlocking loops formed from the ends of the links. Also made by connecting wire links together by a metal object such as a connector, knot, button, tappet, etc. It can also be a continuous wire form.

Check -- A term used in agriculture meaning to plant in check-rows. Also the point of intersection of lines made by a corn-sled. Those intersections are created by marking a field in parallel lines that are spaced to suit the distance between rows, then repeating the process at right angles to the parallel lines.

Check-Line -- The cord, rope, wire, chain, or metal strip on which marks or enlargements are affixed at regular intervals. The check-line is used to determine the distance between rows in check-row planting.

Check-Row -- Rows that are planted in squares so that a cultivator can then operate between rows in cross direction, or the line of checks which define the cross-rows, or the process of planting in check-rows.

Check-Row Planter -- A planting implement that plants in check-row either by mechanical means or by the use of a check-line.

Check-Rower -- A word sometimes used in patents that means the same as check-row planter.

Cord -- A check-line made of cordage, usually cotton, used in the beginning period of check-row planting.

Corn Planter -- An agricultural implement that is designed specifically for planting corn.

Cross-Row -- The early term for check-row, a row that crosses other rows at right angles.

Enlargement -- The object on or in a check-line that is larger than the check-line itself: button, knot, lug, projection, joint, tappet, or stop.

Joint -- The joined ends of the chain links in a check-line or a separate object that is used to join the links such as a metal knot.

Knot -- A knot tied in a rope or cord as was the case with the earliest check-lines. Also the generic term used by collectors to describe any enlargement in any check-line.

Knot Spacing -- The distance between the knots on a check-line, usually one knot per row. In some instances there may be one knot for every two rows or two or more knots per row. Check-lines can exist with knots spaced as far as 96 inches to less than 24 inches apart depending on the corn planter design. Check-lines used for cotton, potato and other plantings may differ.

Lever -- The wood or metal pivot bar that trips the seed drop mechanism. The lever is operated either by hand or forced by a check-line knot.

Link -- A section of wire check-line equal in length to the distance between rows. It may also be the connector acting as a splice that joins two links.

Mechanical Check-Row Planter -- A check-row planter that derives the seed drop action from a series of gears or pulleys without the use of a check-line, sometimes called a "wireless" check-row planter.

Patent -- A monopoly granted by the U.S. patent system to an individual for his invention. In the early times the patent life was 14 years. Later the life was extended to 17 years. Today it is 20 years from the date of application. Most check-line patents were issued during the 17-year ruling.

Planter-Wire -- A generic term used to describe all forms and materials of check-row check-lines and their related knots.

Reel -- A spool-like device on which the check-line is wound.

Re-issue Patent -- A later version of a patent where ownership may be assigned to others, or the text is expanded, or figures are clarified, or an element is separated for exclusive coverage. The duration of the re-issue patent is for whatever time remains of the original patent. The original patent is surrendered.

Rope -- An early medium for the check-line that was used more extensively than cord.

Row Spacing -- The distance between rows, usually 42 to 48 inches for corn cultivation in check-row planting.

Seed Dropper -- The mechanism on a hand or machine planter that released the seed when activated by the linkage or the check-line knot.

Sled -- The name applied to the original cross-row marking implement. The sled was used thereafter by inventors as the basic platform to which all the evolving new features were added.

Stop -- Originally the metal pieces applied to a rope check-line on both sides of the knots to protect the knots from wear and to provide more reliable contact with the seed drop lever.

Throw-off -- A knot or disk on the check-line that served as a de-railer from the guide pulleys to drop the check-line from the planter as it reached the end of a row and prepared to turn around. Its use allowed the farmer to turn the planter around at the end of the check-row without first dismounting. It was used only on planters that ran a straight check-line path.

Wire -- A metal strand of check-line or the wire core within a rope check-line.

NOTE -- The generic term "knot" is used for all enlargements on a check-line. The term "planter wire" is used by collectors for all check-lines regardless of material or knot forms unless otherwise noted.

SECTION 1

HISTORY OF
CHECK-ROW PLANTER WIRES

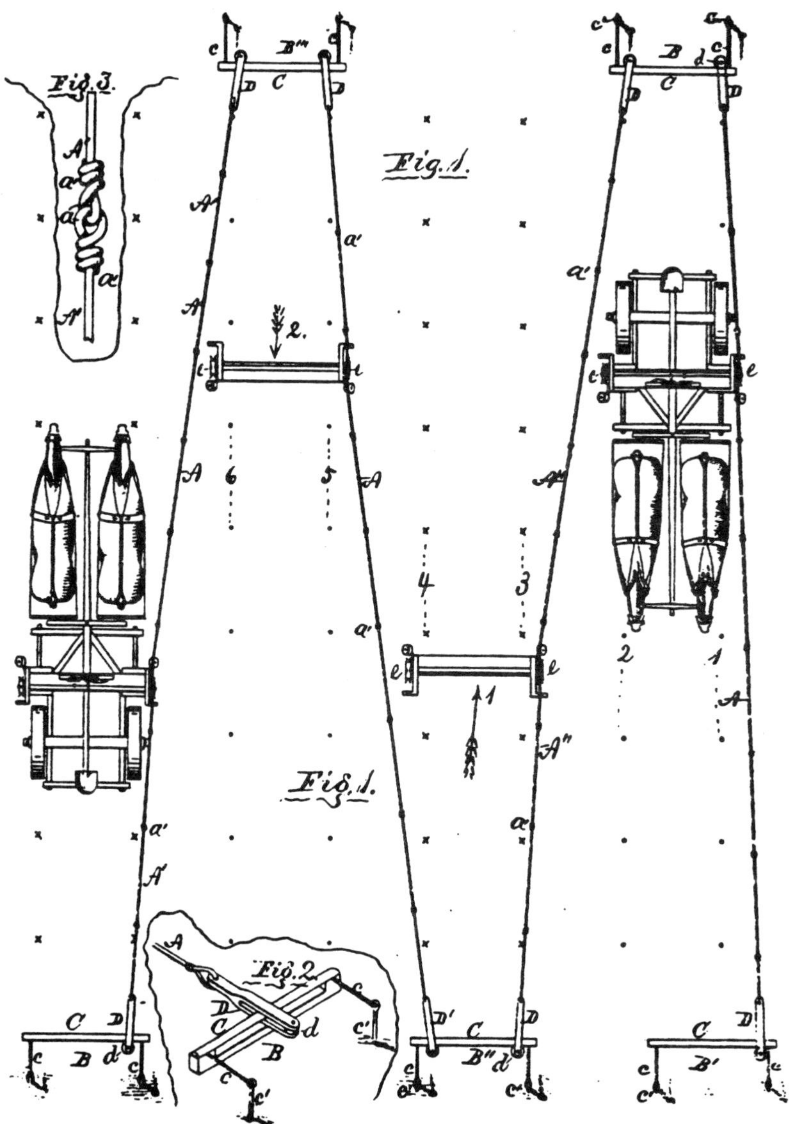

Check-row planting, sometimes called cross-row, refers to crossed rows that are laid out in squares and only planted at the intersections of the squares. The purpose is to allow room for cultivation in cross directions. The term check-row first appeared in the U.S. patent system in the 1850 agricultural section of the Patent Report. There was no Department of Agriculture to champion the farmer in 1850. The Patent Office dealt with all such matters by granting patents for land development, seed improvement, animal husbandry, and mechanics. The agricultural branch of the Patent Office sought any and all information from individuals and universities in order to establish the best farming methods and make them available to everyone.

The Commissioner of Patents in preparing the 1850 Agricultural Report sent out a call to farmers and universities for information as to the nature of the land, type of crops, and the methods of farming that provided the best results in their area. Testimonials from those responding spoke of their great reliance on providence and of the enormous effort necessary to prepare the land for a successful crop.

Farm machinery as we know it today was non-existent. The basic sources of power were man and animal. The crops that resulted were consumed primarily by the farmer, his family, and his animals. Very little was left over for market. Many farmers of the day had little or no experience in farming. Some were recent emigrants who brought with them the farming techniques of the Old World and soon found out they did not necessarily apply. Others were eastern Americans seeking a new life during the period of western expansion. Books or manuals on agriculture were scarce, and those available lacked information that covered the soil types and climatic conditions to be found in the new land. Nor were there enough experienced farmers to supply advice.

Larger farms were now possible. However, with a large farm one needed new methods, different tools, and specialized equipment if success with limited manpower were to be possible; farmers soon became experimenters. The earliest patent granted suggesting a method for laying out rows of corn was issued in 1842 to Peter Moseley of Yazoo County, Mississippi. His patent 2,882 (see page 19) provided for a crude means of surveying parallel corn rows on flat and undulating land.

Most of the testimonials entered in the 1850 Agricultural Report came from farmers who for the first time had the use of more land than was necessary to support themselves. They also said that corn required much hand preparation and cultivation. Land was abundant, but manpower and horsepower were limited, and what was successful in one soil often failed in another.

Testimonials varied regarding the order of cultivation of corn that gave the best results, but they had one common element. Most farmers remarked that it was necessary to space the rows and hills 42 to 48 inches apart for the best cultivation. That width provided the minimum room necessary to allow for horse, man and hoe-mounted sled to work between the plants without causing damage to the roots or plants. In order to space the rows and hills consistently in the 42 to 48 inch space, the farmer needed some way to lay out rows and mark hills in the field. It was common practice to mark the hills by using a wooden sled with runners spaced as far apart as the desired rows. The

(continued on page 20)

December 12, 1842   2,882

"Laying Off Corn-Rows, &c."

"Be it known that I, Peter Moseley, of Yazoo County and State of Mississippi, have invented a new and improved mode of laying off rows for farmers to plant on, say, corn, cotton, or any other kind of thing on a level or to give them any descent that may be desirable on broken, rolling, or undulating lands: and I do hereby declare that the following is a full and exact description."

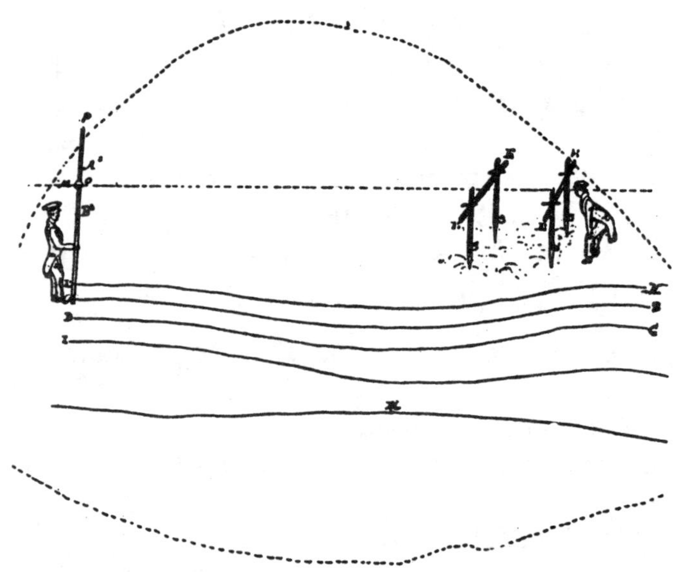

This is the first patent suggesting a process leading to check-row planting. As is the case with new ideas, the attempt seems awkward. The process involved a crude but effective survey to layout rows parallel to each other while allowing for erosion control.

sled was dragged across the field. On the return trip, the farmer used the previous track as a guide, and dragged the sled back, marking the next row. This procedure was repeated until the entire field was marked with parallel lines. Then the process was duplicated at right angles to the parallel lines previously made creating a checker-board pattern. The intersections where the tracks crossed marked the sites where seeds would be planted.

Seeding was done with a walk along hand planter at first. Then over a period of ten or so years, one and later two seed boxes were added to the sled along with furrow opening plows and wheels to close the furrows. A seat for the farmer was added above the wheels as the sled became the basic platform for an evolving corn planting implement. Out of necessity, the farmer had become an inventor and the row-marking sled became a multi-function corn planter.

A common problem experienced by many farmers was the difficulty in controlling seed drop. The farmer struggled with a horse or team, tried to keep the planter in line, and operated the seed drop at each intersection. As a result many hills of corn could be out of alignment and subject to damage during cross-cultivation. Some of these problems were overcome when a seat was added so a helper could operate the seed drop, if a helper could be found.

There is no record of when a rope or cord was first used to show when to drop the seeds. Someone figured that a smooth rope with marks or knots tied on it, then stretched across a field to guide the planter would be a signal to drop the seeds. This eliminated the need to mark off the field in squares before planting. The further addition of an out-rigger row marker along with the marked rope made laying out rows in a field prior to planting unnecessary. The process was simplified, but accuracy was still lacking. If something to automatically trip the seed dropper to drop seed in time with the movement of the planter sled was found, accuracy too would be improved.

The first evidence for such a device was recorded in 1857 with the issuance of patent 16,611, (see page 22), for a corn planter to Martin Robbins of Hamilton, Ohio. He fashioned a chain out of heavy wire in lengths equal to the row spacing. For the link connector he used a metal button with wire eyes extending from both faces. Hooks were formed on each end of the links and attached to the eyes of the button, creating a continuous chain. Next he mounted an attachment on the planter sled which supported the chain by means of guides, allowing the button to contact the trip lever. The chain was staked at one end of the field and stretched to the other side. A horse pulled the planter alongside the chain. Now, as they proceeded across the field, the seed box tripped when the lever contacted the buttons. Robbins had eliminated the need for a helper on the planter. One man alone could plant a field and with better accuracy. This was a giant step in planter development. The use of a check-line for check-row planting was a reality, but it would be a few more years before the idea took hold.

About the same time, others developed check-lines using knotted ropes instead of chains. That was the natural next step for those already using a knotted rope as a visual marker. They probably continued using a knotted rope because it was easier to make it suit individual row spacing and rope was readily available. The first patent disclosing the use of a knotted rope as a check-line on a planter was patent 44,472 issued to John Thomson and John Ramsey of

(continued on page 23)

February 10, 1857

16,611

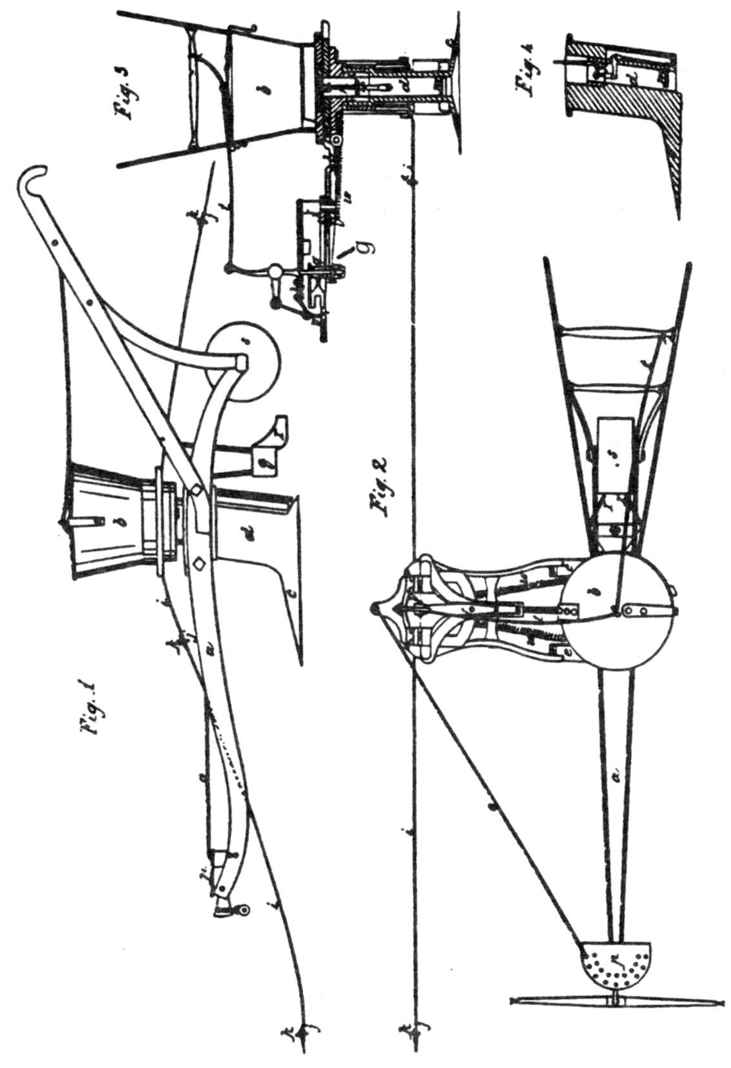

History of Check-Row Planter Wires

Aledo, Illinois, September 24,1864. The rope was wound on a reel mounted on the planter. The end of the rope was anchored at one edge of the field and unwound as the planter made its trip to the other side. Upon reaching the end of the row someone had to go back to release the rope so it could be rewound. Rewinding the reel was done manually with a crank.

By 1870 check-line attachments were being made to adapt to existing planters. G.D. Haworth of Decatur, Illinois obtained patent 100,032. For the first such attachment, he used a knotted rope. He mounted this attachment to his planter, patent 44,725 which he had previously obtained in 1864. His design routed the check-line across the planter. This was intended to decrease side stress on the horse and disruption to the planter and check-line.

It was nearly impossible to tie knots evenly spaced in a rope; the knots wore, they did not always trip the lever, and eventually the rope broke. To overcome these problems, William Hubbard came up with an idea to add stops and balls to the rope. He was issued patent 54,166 in 1866 for his effort. Stops or balls could be secured to a rope like stringing beads. There were still problems with rope, but the invention was significant. Hubbard's patent was the first to mention balls or stops applied to a rope, although, no specific description of them was cited. Haworth did not show balls or stops in his 1870 patent because it would have been an infringement on Hubbard's patent. Haworth solved the issue later with the purchase of Hubbard's patent, re-issue 7,577. William Grimes was issued patent 113,761 on April 18, 1871, for a corn planter with a seed drop driven by a smooth rope, eliminating the "troublesome knotted rope" check-line. In his patent, Grimes refers to an earlier patent in which he used a knotted rope. Patent 113,761 also shows reference figures from that earlier patent which illustrated metal knots and stops on a rope in lieu of the traditional tied rope knot. Unfortunately, Grimes did not identify his earlier patent and research has been unable to locate it.

In spite of all efforts to improve rope check-line, rope still stretched and broke, and when the rope got wet, it shrank. To overcome this some patentees reinforced the rope by adding a wire core. This was an improvement but still the rope was subject to rot and mold. It held dirt which clogged the pulleys and it became a choice material for nesting rodents. Rope wore out prematurely at the knot due to impact with the lever. Rope check-lines were never as durable or accurate as chain or wire. However, in spite of these deficiencies the knotted rope check-line continued in general use for some time. In an effort to remedy this, the next patented improvement was to add a one-piece metallic stop to both sides of the knot. On September 19,1871, Lysander Haworth was issued patent 119,142 specifically for a metal stop that straddled a knot or a splice on a cord to lessen wear. The patent improvement was called a stop, however, the term knot continued throughout check-row usage.

While some inventive farmers were working to improve check-lines, others were trying equally as hard to devise a mechanical method to replace the need for a check-line and still achieve the same results. Some relied on wheel traction to operate the seed drop. Others tried using a smooth rope or wire tied across the field, much the same as a check-line, but trapped around a driver pulley on the planter to friction drive the drop mechanism.

In 1869 S.Y. Orr was issued patent 91,961 for such a planter, whereby wire or rope was wound on reels mounted directly on each side of the planter. One end of the wire or rope was tied at the edge of the field. On the first pass across the field, the wire or rope unwound from one reel. Controlled drag, caused by a tension clutch on the reel shaft was sufficient to turn the seed drop mechanism. Once across the field the planter was turned around and the wire or rope from the second reel was anchored as before. On the return trip, the first reel rewound and the second reel played out. The process was repeated for the remainder of the field. There is no evidence that this idea ever gained any popularity.

Now that check-lines were becoming more reliable and more planters were designed to accept check-lines inventive competition was on to obtain the best check-rower. Competition continued for the next sixty-five years.

Interestingly, in 1879 the question of narrower row spacing and less cultivation arose. Garitt Hyer of Clinton, Illinois, in his patent 212,469 for a friction drive planter using a rope marked with a piece of rag, string or even baling wire to act as a visual indicator, added the following remark in summation, ". . .prairie country experience is beginning to show that more corn is raised to the acre, and at less cultivation, by drilling the corn in rows running only in one direction instead of planting it in rows at right angles by means of check-row." His comment seemed out of place at the time and he continued with his friction drive planter in spite of its awkward operation. If the seed drop got out of timing, the planter needed to be stopped to allow the farmer to loosen the rope so the pulley could be slipped back into time with it, a troublesome process. While he envisioned the end of check-row planting, he did not make the necessary leap beyond the inventive process of the day which was limited to one improvement at a time.

The search for the best system resulted in well over a hundred varieties of check-lines. Hundreds of planters were patented, but most used some form of existing check-line or a variation. The planter itself became of primary importance now, and the check-line was commonly shown in patent figures with little or no description. Most patents simply said ". . . and check-line now in common usage."

The use of the check-line in check-row planting has nearly died out. The last known patent for a check-line was issued in 1934 when patent 1,982,434 was granted to Arnold Johnson for an improvement over a similar design by Earl Graves in his patent 1,982,427. Both patents were assigned to the International Harvester Company.

In a recent interview, Gerald Huebert, a retired farmer near Portsmouth, Iowa, recalled using wire check-line for planting. When he was only eight years old his father started him out on a two-row check-row planter. His father "checked corn" with an old horse-drawn planter until he retired from farming. Mr. Huebert used a John Deere no. 999 check-rower, also horse drawn, until 1964. He planted 100 acres that year. The check-line had knots similar to the Barlow or Parker check-line with a forty-inch knot spacing. Mr. Huebert never gave much thought to the type of wire, but he could not remember ever having it break. "It was a heavy, strong wire." After 1964, he used a tractor and a front mounted hilldrop planter. "With that planter I could see what was going on. It was a lot easier," he said.

Check-row planting has been replaced with the more efficient drill planter. The plants grow close together in narrower rows in one direction. Hybrid seed, chemical fertilizers, herbicides and pesticides have greatly reduced, and in some cases, eliminated the need for cultivation after planting. This is perhaps what Garitt Hyer had envisioned in 1879 as he struggled along with his rope check-line. He was just a little ahead of his time.

Today check-row planting with check-line can sometimes be seen at shows and fairs where old equipment is being demonstrated. Some small farms and some worked by various ethnic farmers are also known to still use this method. For most farmers it is too time costly for practical commerce, and except for the concern and efforts of the collector, the check-line would no doubt disappear forever.

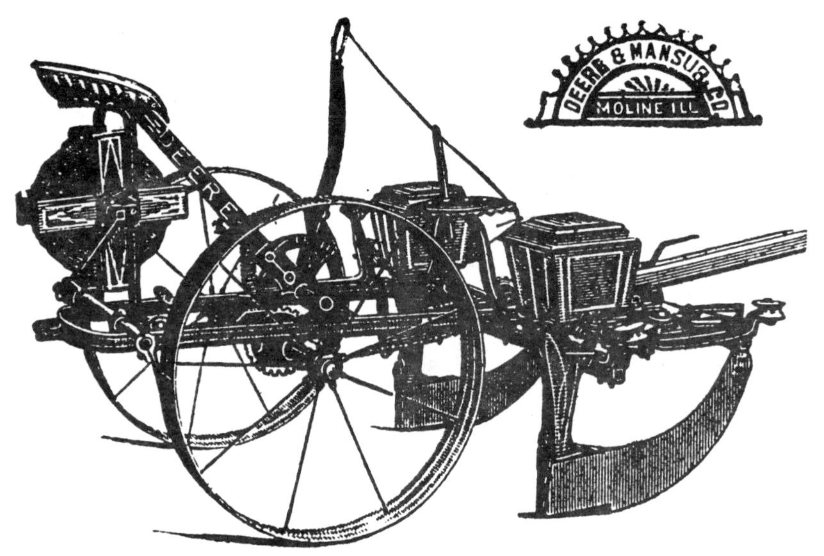

DEERE STEEL FRAME PLANTER.

**From a Deere & Mansur, Co. envelope
postmarked July 28, 1893**

RETURN IN FIVE DAYS
Deere & Mansur Company
MOLINE
DEERE STEEL FRAME PLANTER.

The Barlow Corn Planter

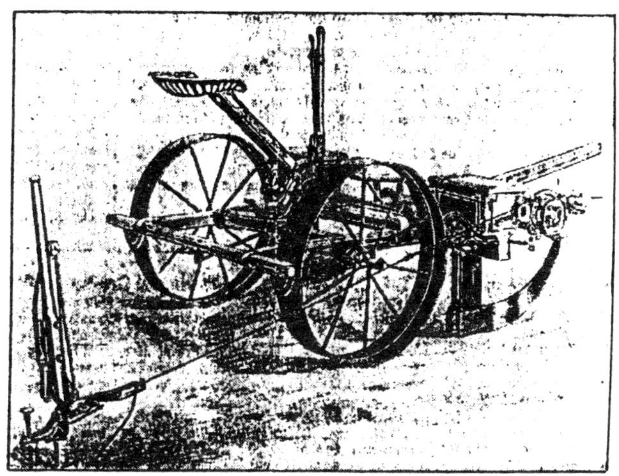

The Barlow corn planter, invented by Joseph C. Barlow after the Civil War. It won awards at the Centennial Exposition in Philadelphia in 1876. One feature was its ability to seed three corn hills at a time.

Photo courtesy of Carl Landrum and the Quincy Herald Whig, Quincy, Illinois.

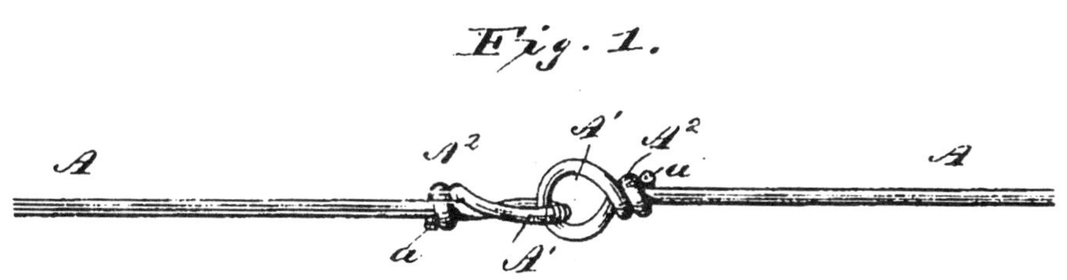

The above picture is the Barlow corn planter displayed at the Centennial Exposition in Philadelphia in 1876. The check-line is the early style shown in Fig. 1 of his patent 328,452. The anchor is 373,170 patented in 1887. This suggests that the picture for the article was taken later for the publication.

SECTION 2

CHECK-LINES AND KNOTS

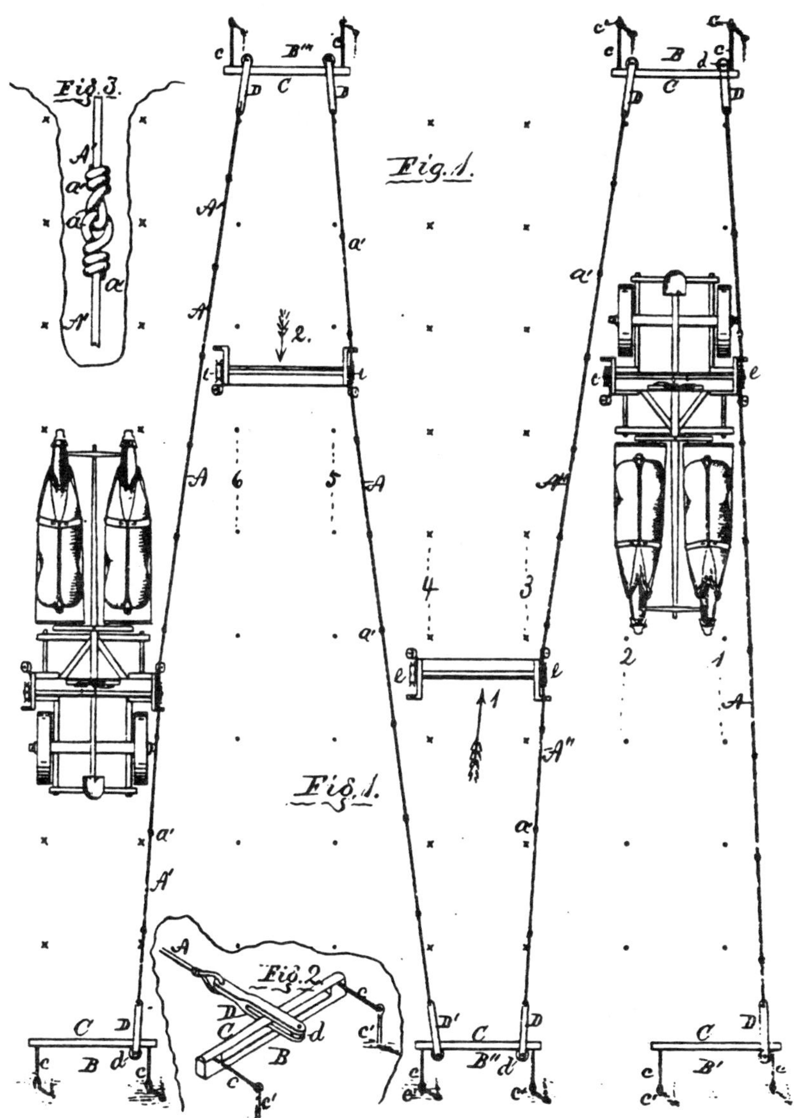

MARTIN ROBBINS

February 10, 1857   16,611

This is the first U.S. patent on Check-row depicting the check-line and related knot.

"Be it known that I, Martin Robbins, of Cincinnati, Hamilton County, Ohio have invented a new and useful improvement in corn planters. The object of my invention is to produce a seed-planting plow which (without previous laying off the ground) will plant in equidistant and opposite hills, and thus admit of cross plowing and cultivation."

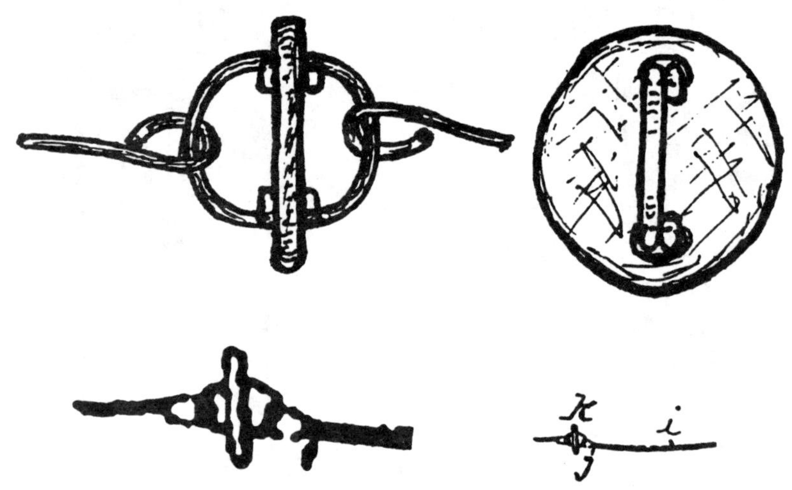

"Chain i,j,k of which the rods i correspond in length to the desired distance between the consecutive hills. These rods are united end to end by means of hook-and-eye joints j. At each joint is a button k."

"I claim. . ., in combination with the jointed rod or chain provided with buttons or similar devices, for the purpose explained."

MARTIN ROBBINS

February 9, 1858   Button   Re-issue 525

Exactly one year from Robbin's patent 16,611 he obtained a Re-issue specifying the check-line in more detail, and stated specifically that the purpose was for 'check-rows.'

"The object of my invention is to enable a seed-planting plow, without any previous 'laying off' of the ground, to plant in equidistant and opposite hills or 'check-rows,' so as to admit of cross plowing and cultivation.

The device chiefly instrumental in effecting the regular automatic planting in check-rows consists as follows: ijk is a chain composed of rods i, united end to end by equidistant hook-and-eye joints, j, each joint being provided with a button, k, which is the immediate means of vibrating the tappet g (see fig. 3, pg. 22). These buttons are of course placed at a distance apart corresponding with that desired between the consecutive hills.

The particular method described of constructing the chain is selected, because it has been approved in actual use, and because it illustrates all the material features of that part of the invention; . . .

The hooks and eyes enable the chain to be lengthened or shortened to suit any field, and facilitate its extraction in case of its becoming entangled. Common snap-links may be substituted for the said hook-and-eye joints, if preferred. For example, for the better class of machines a fine galvanized wire cord may take the place of the jointed rod or chain, which the cord may be wound on a reel attached to the machine, and one end being anchored it may be laid down in the act of planting the first row of hills."

February 9, 1858   Button   Re-issue 525

"Common snap-links may be substituted for the said hook-and-eye joints, if preferred."

The below sketch depicts the option of using snap-links. The button likely is the same as that described in 16,611. A galvanized wire-cord may also be used.

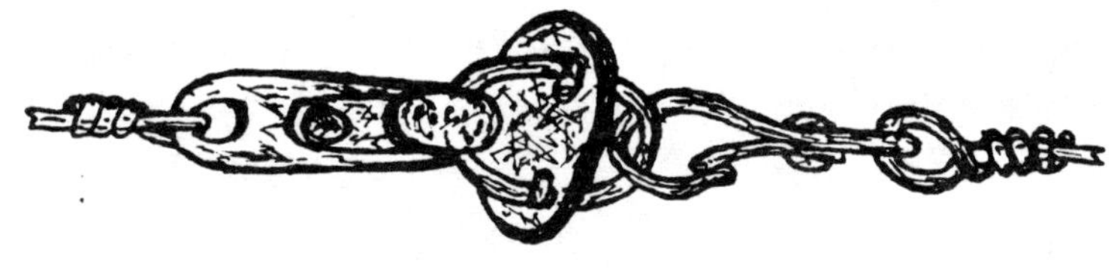

The mounting of this check-line in reels on the planter is the first, and differs from the statement in 91,961, later, where a smooth check-line is in reels and mounted on the planter.

September 24, 1864   Knot   44,472

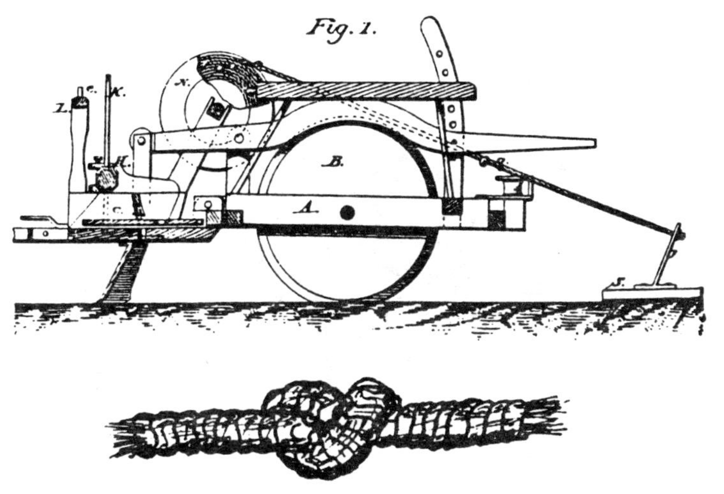

Aledo, Mercer County, Illinois

"The employment or use of a wire or cord *o* provided with knots at a suitable distance apart and applied to the machine substantially as shown, in connection with anchors *s* all arranged as and for the purpose set forth."

This patent is the first to use the knotted cord, and became basic to all the Haworth patents to follow. The knot was simply tied in the check cord.

The Haworth family purchased this patent to protect their growing business patents. The first re-issue 6,821 on December 21, 1875 claimed: "The combination with a corn planter of a <u>knotted cord</u> for actuating the machine." The second re-issue 7,378 on October 31, 1876 claimed: "In combination with the <u>cord, or its equivalent</u>, actuating the seeding devices."

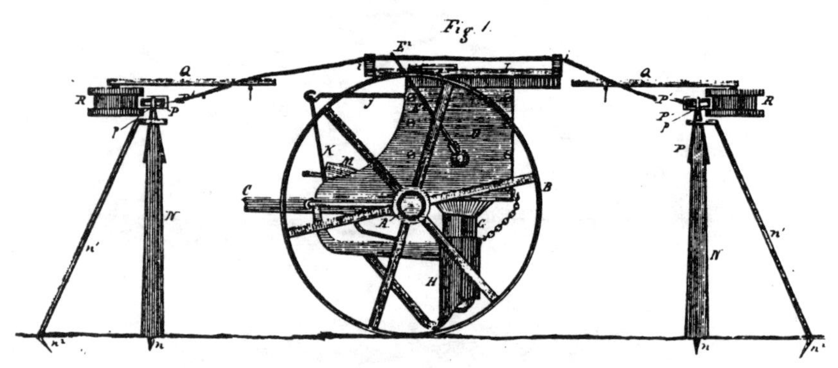

Edinburg, Johnson County, Indiana

This patent is the first to mention the use of balls and stops. Though no mention of the term check-row or cross-check is made, this indeed is such a planter. The patent figures do not show any detail of the wire/cord or balls and stops.

"I stretch a cord or wire, and upon this cord or wire are balls or stops permanently secured every three or four feet apart or at any desired distance."

The Haworth family recognized the significance of this patent since it is basic to the stated balls and stops. They purchased the patent to protect their other patent interests, much as the Washburn & Moen Company did in the barbed wire industry. Thus, it was re-issued to George, James, Lysander, and Mahlon Haworth on March 27, 1877 as RE-7,577.

S.Y. ORR

June 22, 1869   Smooth Check Wire/Cord   91,961

"This invention relates to a new and improved planter of that class which is designed for planting in check-rows, ... "

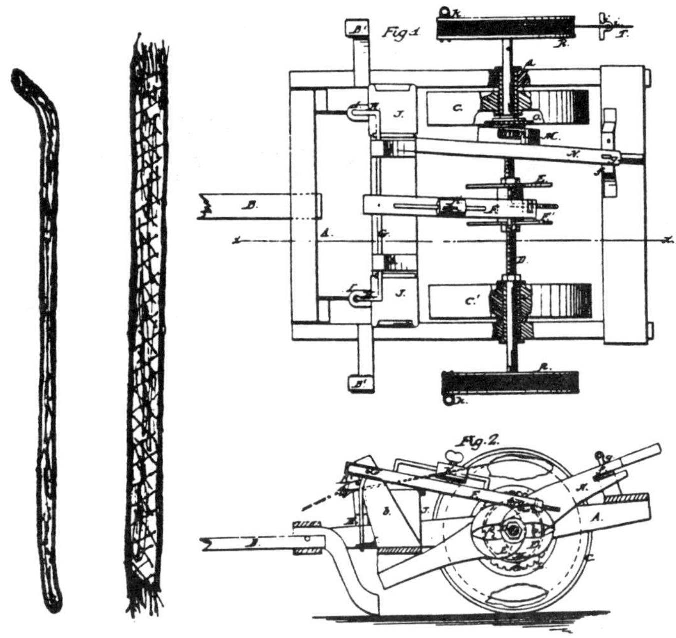

Morning Sun, Louisa County, Iowa

This is the first patent where the check-line is reel mounted on the planter, and a wire or cord turns the shaft that actuates the seed drop. Knots are not used.

"On each end of the axle D there is keyed a reel R having a fine wire or cord S wound upon them. These wires and reels are for the purpose of turning the axle D and operating the seed distributing mechanism as the machine is drawn along."

G.D. HAWORTH

February 22, 1870   100,032

This is the first patent to specify the use of a double-spaced knotted cord. 44,472 had already claimed the knotted cord. Haworth later acquired that patent in 1876 by the re-issue 7,378. This became the foundation for all following Haworth check-line patents.

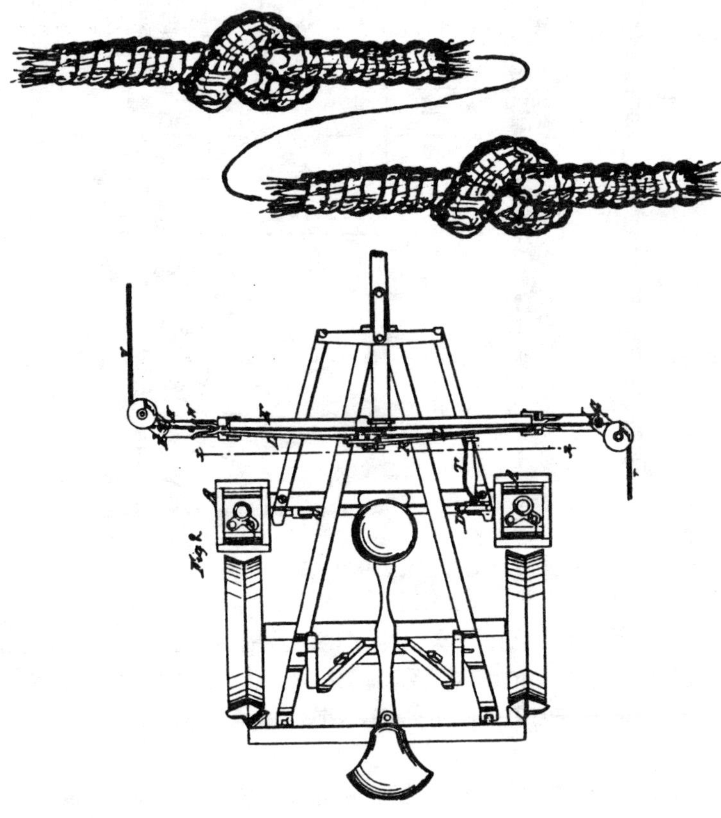

Decatur, Macon County, Illinois

"The knots are made in the cord twice the distances apart which the hills of corn are desired to be."

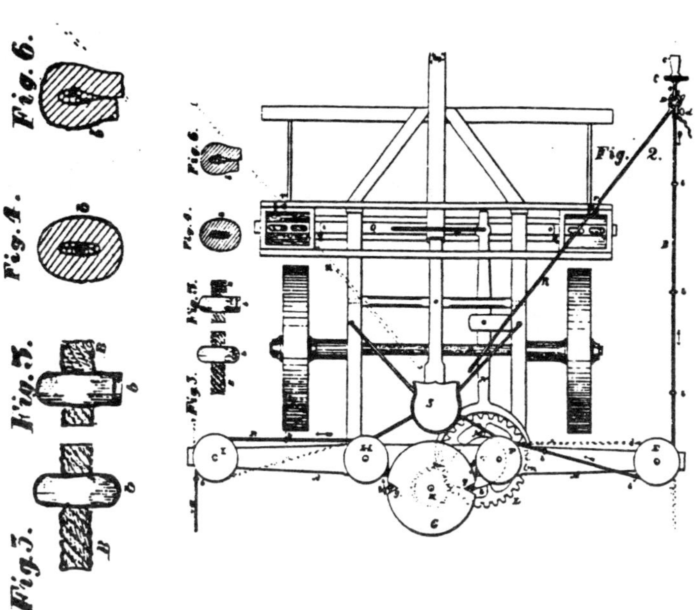

Decatur, Macon County, Illinois

This patent is for an improvement to an earlier one by Grimes and provides for a friction drive eliminating the usual knotted cord. The above figures are of that patent revealed by Grimes showing knots as reference. Two types of knots are shown, Figs. 5 and 6 are a half-round, U-shaped wire or strip pinched down on the cord. Figs. 3 and 4 show a doughnut shaped knot slid onto the cord and pinched.

The figures are shown here so that you will be aware of other types of knots by Grimes that likely existed before the date of this patent.

LYSANDER L. HAWORTH

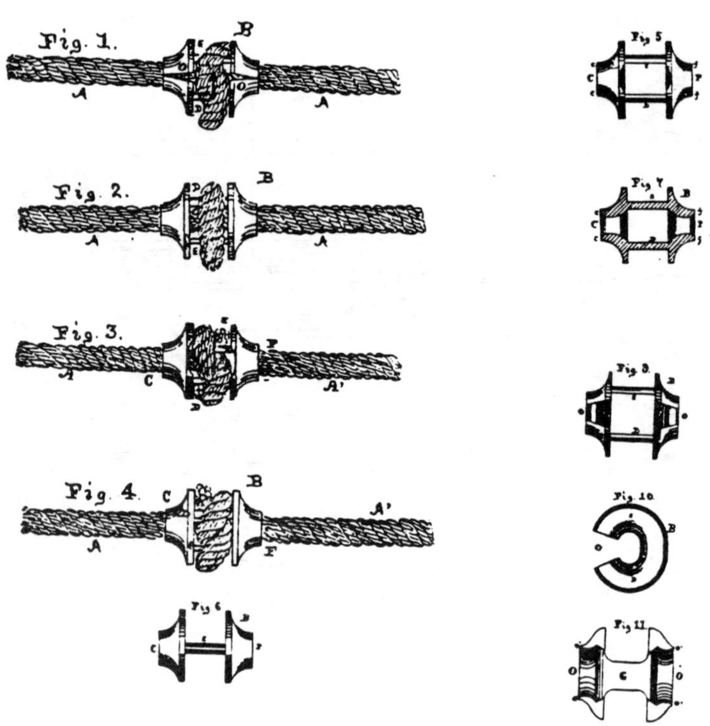

Decatur, Macon County, Illinois

This stop is an improvement on the knotted cord that is used to operate the seed dropping mechanism of a corn planter patented on September 27, 1864, by John Thomson and John Ramsey, No. 44,472, and of a corn planter patented February 22, 1870, by George D. Haworth, No. 100,032.

This patent is the one referred to in the Haworth patent 155,024. "By means of the stop, the cord is prevented from wearing at the knots, as it does when only a knot in a cord is used."

JOHN THOMSON

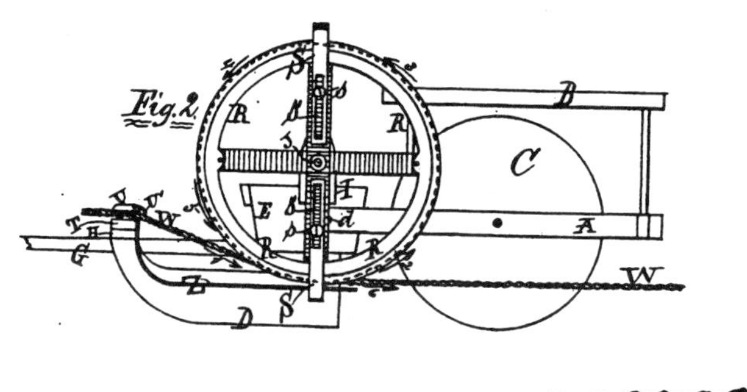

Aledo, Mercer County, Illinois

". . . and the invention consists in an arrangement of the droppers, devices for operating, and in the arrangement of guides or witnesses on the pulleys, by which the operator may see where the hills are being placed, ..."

The cord W being first drawn across the field and its ends made fast, it is then placed across the forward end of the machine, and over the two pulleys UU as shown plainly at Fig. 1. It will now be seen that in advancing the machine forward, the cord will rotate the pulley R and operate the dropper twice to each revolution of the pulley. " . . . and when from any cause, the hills are not dropped exactly in line, crosswise with the line of progression of the machine, the driver may stop the draft animals and turn the pulley within the cord until the indicators $S\,S$, which have shown the error, indicate the correction by pointing or standing in a vertical position opposite to the mark made by a projecting of one in the last rows planted."

November 5, 1872   Knotted Chain   132,792

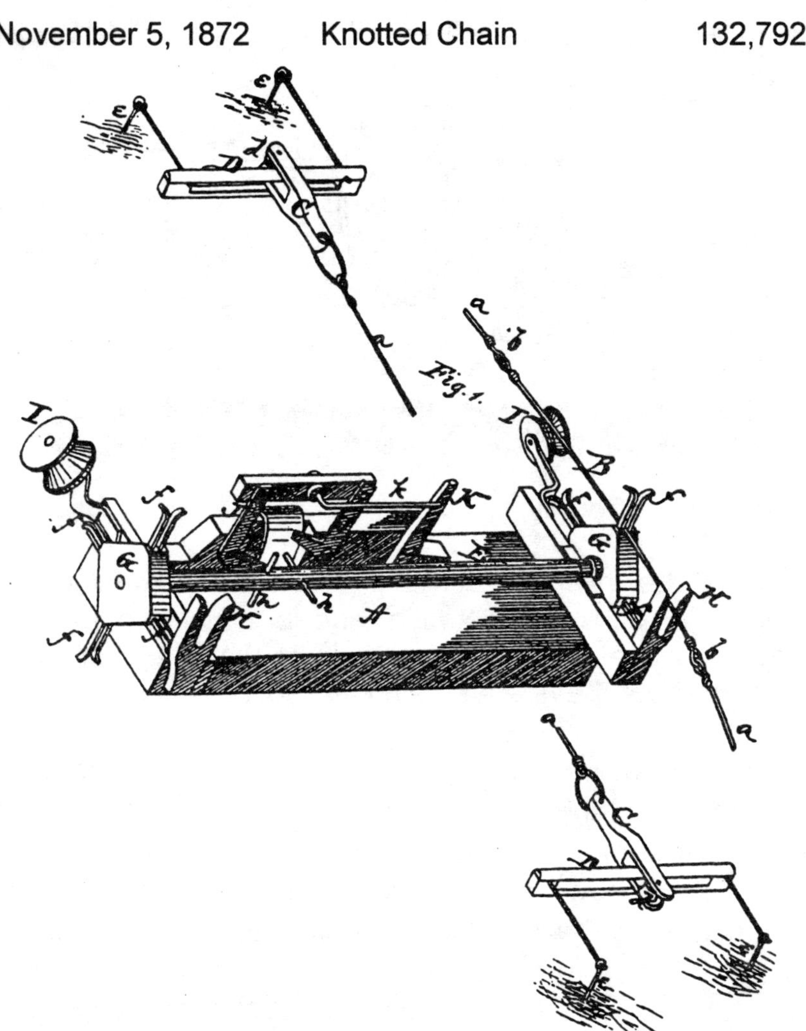

Bloomington, McLean County, Illinois

This is for a complete check-row attachment including the chain shown in Fig. 1. This patent re-issued in 1877, re-issue 7,522, separating the chain from the attachment. Another re-issue, RE-7,521, claims the rest.

WILLIAM T.F. SMITH

December 10, 1872   Knob   133,808

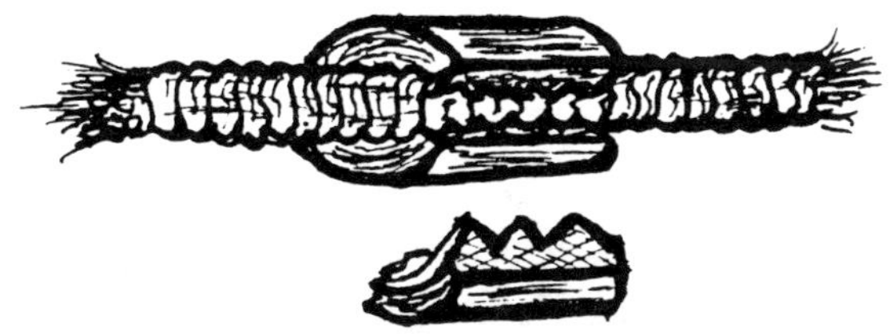

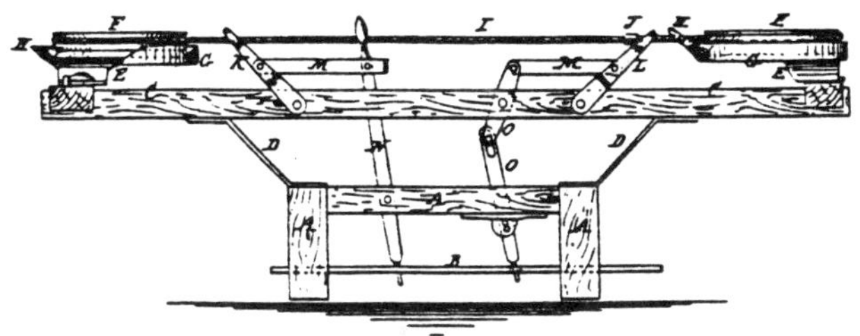

Lexington, McLean County, Illinois

"Upon the rope I at suitable distances apart, are secured metallic knobs, J to operate the levers or tilters, as described. The knobs J are made of bars of some metal of sufficient hardness, which are grooved longitudinally, as shown in Fig. 4, so that when cut into pieces of suitable length and wrapped around the rope they may clamp the rope transversely, and thus be prevented from slipping."

The grooves in the knot are the improvement over Grimes, 113,761.

January 14, 1873   134,747

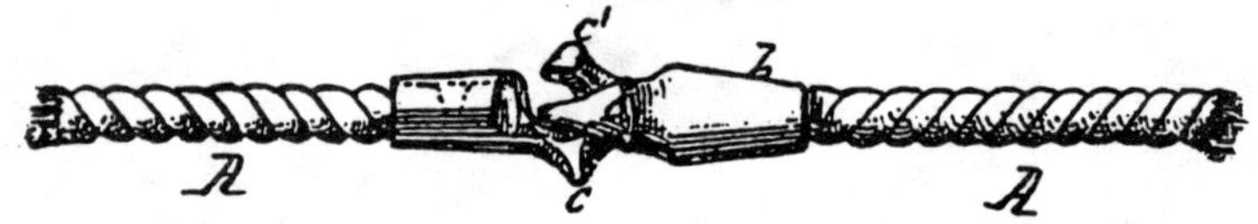

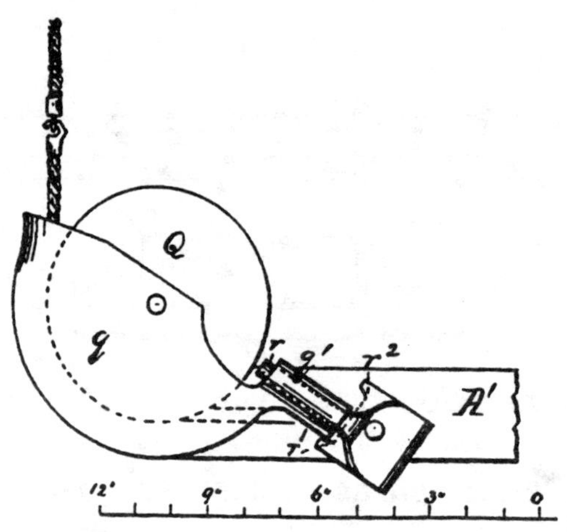

Decatur, Macon County, Illinois

This patent is an improvement on Haworth's 100,032 planter. It is shown here to note the cord with the open hooks that Haworth detailed later in re-issue patent 7,235.

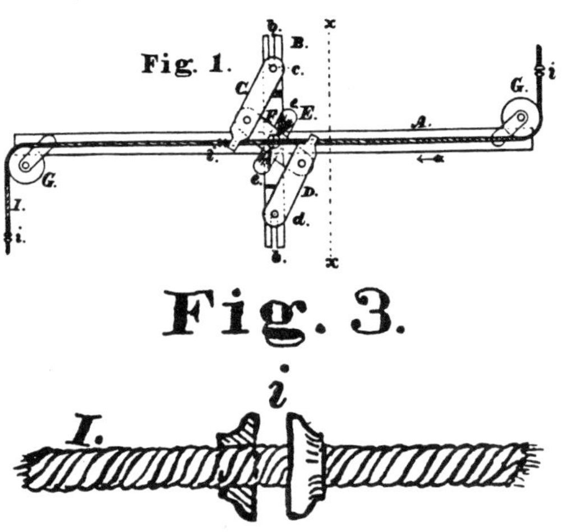

Decatur, Macon County, Illinois

This patent is an improvement on Haworth's 100,032 planter, and specifically for what is called a "one horse-planter." It claims the use of two stops with space between so that the cord will bend freely around the pulleys. Note that the stops are the same as were used to protect the knots earlier. <u>The improvement is the absence of the knot.</u>

"The stop consists of two pieces of malleable or wrought iron i, rolled or formed as shown in Fig. 3, being made open, placed over and compressed upon the Rope I being placed at such a distance apart that they will allow of the rope bending when passing around the pulleys, and also, as the arms only strike on the one or outside of the stop, it will not work backward and forward, chafing and wearing the rope."

BENJAMIN B. HILL

June 2, 1874   Cord   151,590

Fig. 1.

Worcester, Worcester County, Massachusetts

"The nature of my invention consists in a cord composed of metallic wire and fibrous threads twisted together to make up the strands, to be twisted and then laid together in the same manner that ordinary cord is twisted and laid."

"This method of making cordage containing metallic wire enables me to produce a nice flexible line or cord, very strong and durable for various uses. Twisting the wire in connection with the threads holds the thread so firmly that it cannot slip on the wire."

This patent very likely became part of the Washburn & Moen Company, and found its way into some of the planter wire business. Several planter wire patents specified such a cord. The number of strands was not a specification of the patent.

July 28, 1874   Chain   153,576

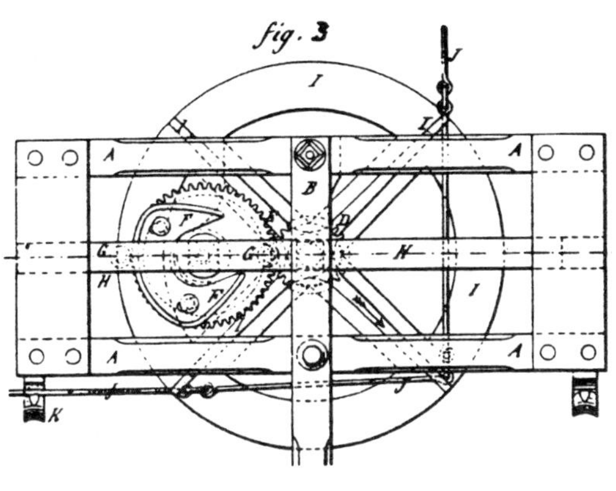

Camp Point, Adams County, Illinois

"The chain *J* is made of long links, of a length equal to their distance between the notched ends of the arms of the wheel *I* as shown in Fig. 3."

LYSANDER L. HAWORTH

September 15, 1874   Joint   155,024

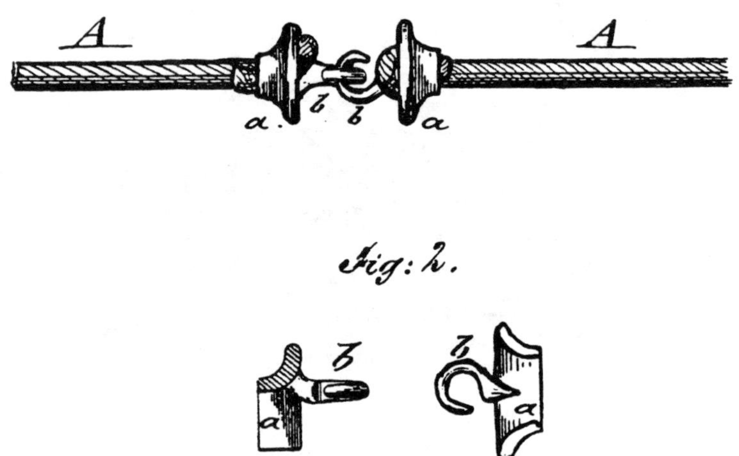

London, Madison County, Ohio

"The object of my invention is to provide a joint for check-row cords used for dropping devices in corn planters, so that the cord can be readily unhooked and passed around trees and be hooked again without requiring the changing of the corn planting implement or the position of the cord across the field."

"My invention consists of a metallic bell-shaped sleeve or ring with projecting hook, which is jointed to the connecting hook, while the sleeve is firmly closed or clinched on the loop-shaped cord end, after passing the same around the hook."

JOHN M. BANKER

November 17, 1874   Knot   156,870

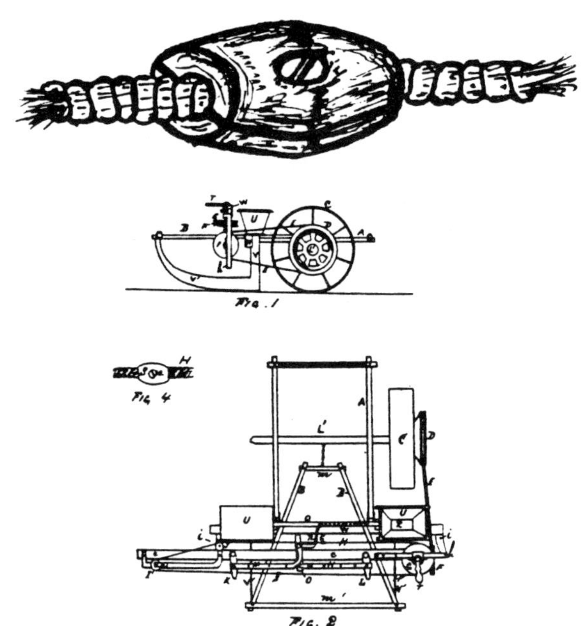

Tuscola, Douglas County, Illinois

This patent makes the first reference to the provision of an "artificial knot" over the traditional hand tied knot in the cord.

"The artificial knot S, Fig. 4, consists of two oblong rounded pieces of wood or iron, having flat faces, grooved longitudinally to receive the rope, being fastened together and to the rope by the screw a'. A knot of this description is much better than one tied in the rope in the usual manner, as it can be readily adjusted to any position, is cheaply constructed, and prevents the rope from wearing out and breaking."

January 26, 1875   Stop   159,177

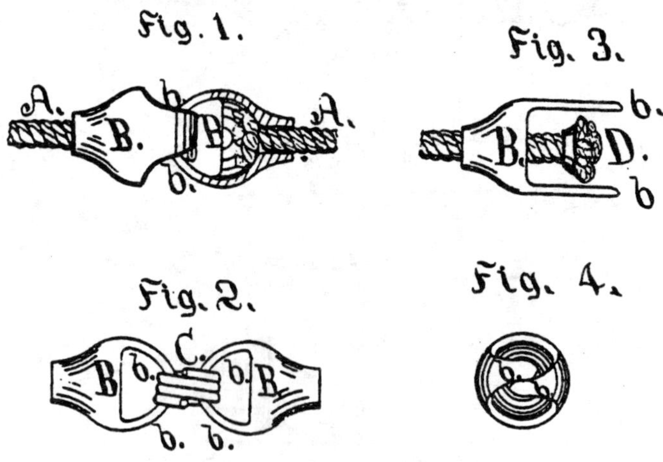

Decatur, Macon County, Illinois

This patent describes three different types of stops.

Fig. 1, is an improved stop forming a universal joint to ease strain on the cord at the pulleys.

Fig. 2, shows the stop with spring connection, acting as an adjustable coupling. These are used every eight to ten rods.

Fig. 3, shows a section of the stop with swivel fitting to allow for torsion relief on the cord. These may also be applied every eight to ten rods.

February 23, 1875   Stop   160,055

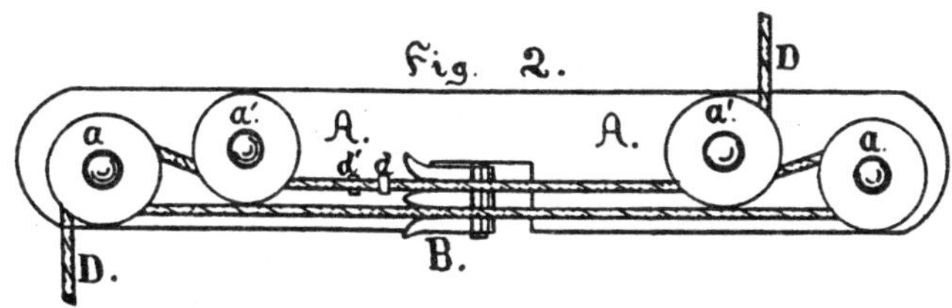

Decatur, Macon County, Illinois

"My invention consists in producing a simplified check-rower, by means to be hereinafter specified, that will operate twice for every stop on the cord, will lessen the side tension or leverage, and can be attached to the planter in the rear of the dropping apparatus, which last arrangement is specifically adapted to what are commonly known as one-horse planters."

This patent is another whereby the space between stops is twice the row distance. Note that the rope changes direction allowing for the use of just one trip lever.

September 7, 1875   Buttons/Knots   167,557

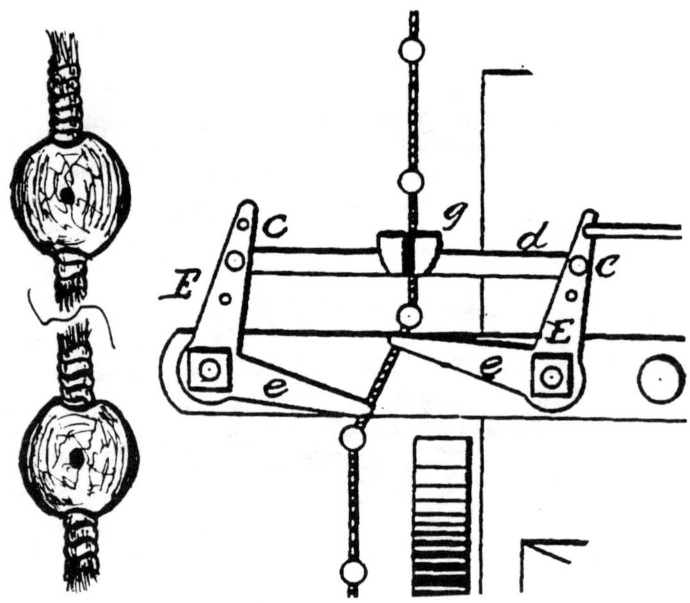

Fairbury, Livingston County, Illinois

This patent is one of several where the check-row attachment was so designed as to use closer knots to operate the levers rather than relying on springs or other devices to return levers to the starting position.

"As the knot or button slips through the eye it will catch in the fork of this angle-lever and pull it to the rear, throwing by means of the Guide-bar d, the other lever with its forked end forward to the eye, g to receive the next knot or button of the rope, and in turn drawn to the rear."

November 5, 1875   Chain   169,865

Aledo, Mercer County, Illinois

"This machine is made to do its check-rowing by means of a flexible chain. The machine runs close to the chain across the field. When it moves the wheel I and the bar G revolve, and Cams aa, operating not only on the seed-slides and upon the markers, not only distribute the grain, but mark where it is deposited."

No specification was given for the chain.

December 7, 1875   Knot   170,673

Farmer City, DeWitt County, Illinois

"Instead of constructing the line of tarred rope, hemp, manila, or such like material, <u>I make it of twisted wire</u>, and instead of using <u>only one set of knots, I use two</u>."

"The object of thus providing two sets of knots is that in ordinary line where only one set is used, in turning around at the end of the field a certain length or space is lost or taken up in the line so that the knots are not in proper position to drop the seed correctly in the reverse direction of the machine, as they were in the first instance. I provide these double series of knots therefore, at the proper position, so as to take up or compensate for this loss, so that in turning around at the end of the field the second knot of the pair will be in proper position, so as to drop the seed correctly."

July 18, 1876   Knobs   Re-Issue 7,235

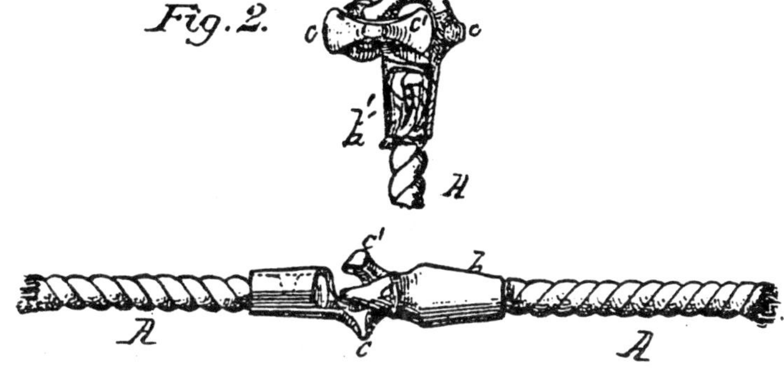

Decatur, Macon County, Illinois

This re-issue extracts the knot from 134,747 where it had been shown, but not detailed.

"In the drawings, the shanks are shown as bent into the form of tubes or sleeves surrounding the end of the cord and broken away in places. Figs. 1 and 2 show a pointed spur *B*, which penetrates the cord or around which the end is looped."

"The hooks readily turn one within the other, forming a universal joint connection when united. Thus formed, the hooks can be united or separated only when brought into a certain angle of relation to each other."

LYSANDER L. HAWORTH

October 3, 1876   Buttons   182,820

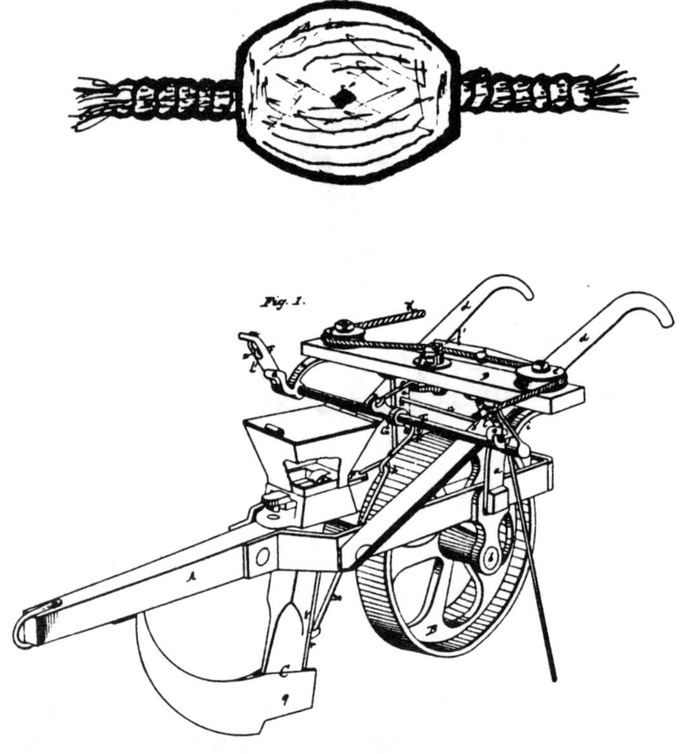

Decatur, Macon County, Illinois

This patent is for a corn planter. The knot shown on the cord appears to be oval in its shape and likely wooden. No description was noted in the patent.

October 10, 1876   Tags   183,113

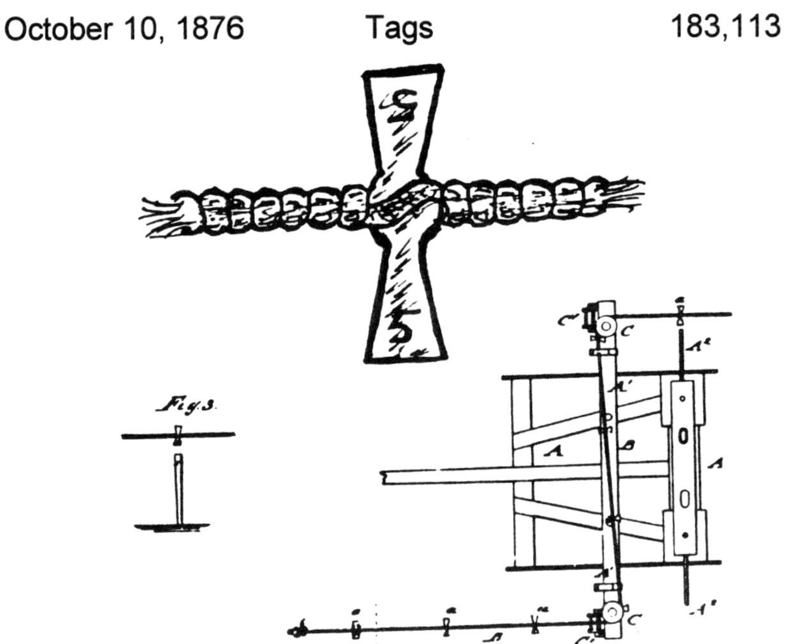

Pawnee, Sangamon County, Illinois

"A vehicle, having Cross-bar A', Pointers A'', Pulleys C, and Guides C', in combination with a cord, B, having tags a arranged for the purpose specified.

"This mode of rowing saves the time of marking off the ground and enables the farmer to plant in freshly broken ground so that the corn has a chance to get up before the weeds."

"The tags and stakes may be numbered so the operator can return to an unfinished field and pick up where he left off so rowing and dropping may be continued in the same regular and even manner as before."

December 5, 1876   Balls   185,092

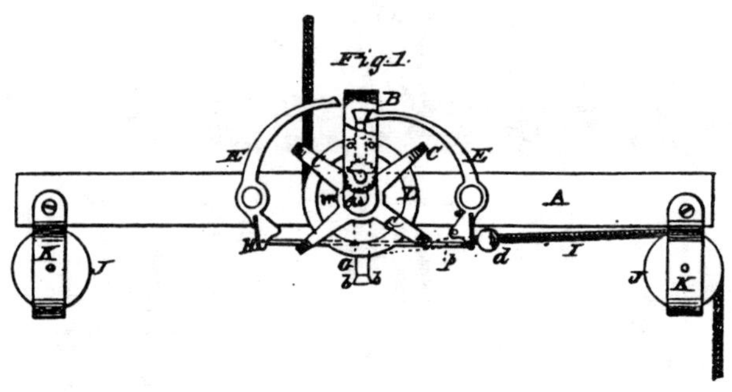

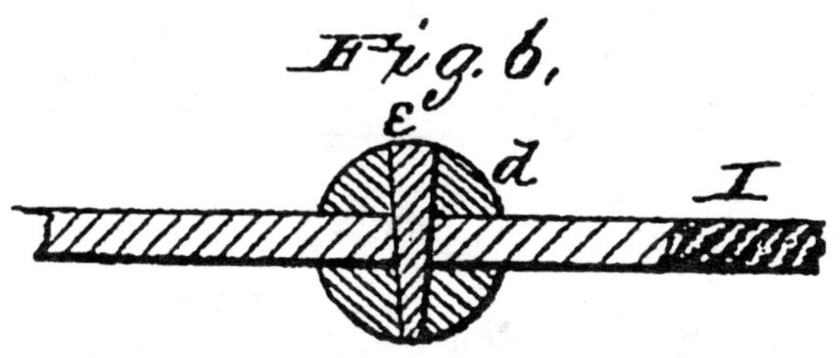

Maroa, Macon County, Illinois

"This rope is, at proper intervals, provided with balls d which are cast around the rope and around pins E passed through the rope previous to casting the balls, by which means it is impossible for the balls to slip on the rope."

Claim 4: The rope is provided with tapering pins E passed through it and the balls d cast around the same for the purpose herein set forth."

ALDEN BARNES

February 20, 1877   Enlargement,   Re-issue 7,522
                      Knot, Trippet

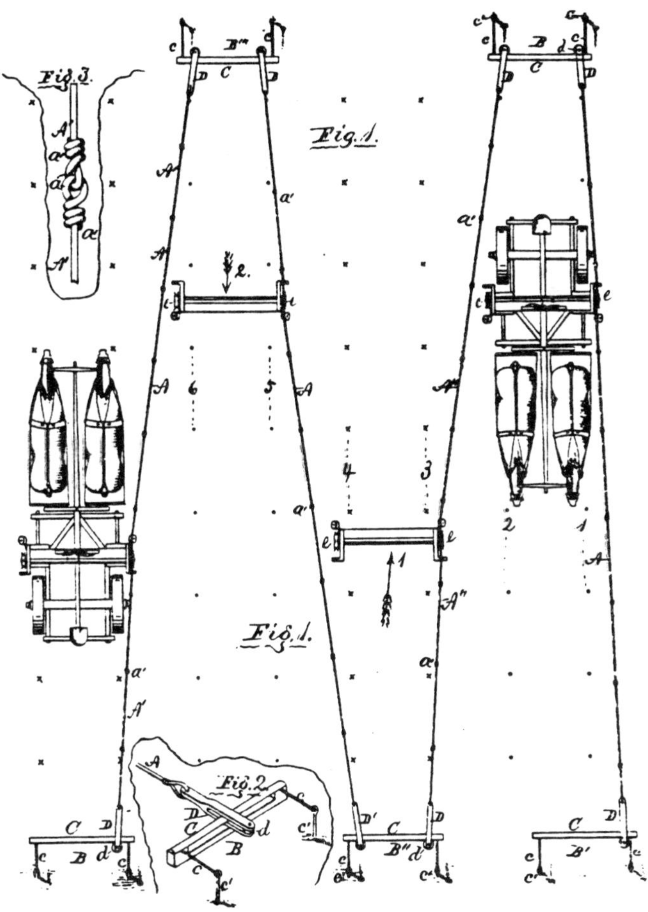

A diagram of the planting process with the Barnes planter
and check-row attachments.

ALDEN BARNES

February 20, 1877	Enlargement, Knot, Trippet	Re-issue 7,522

Bloomington, McLean County, Illinois

This patent separates the check-row chain from his basic patent 132,792 of November 5, 1872. The re-issue claims:

"In combination with a corn planter, a check-row chain having knots a formed by coiling a portion of wire around the main wire, substantially as described, and for the purpose specified."

"Fig. 3 defines the knot. Its design was protected since 1872. This re-issue also specifies the planting process preferred. A second re-issue, 7,521, covered the check-row attachment."

September 18, 1877   Marks and Knots   195,392

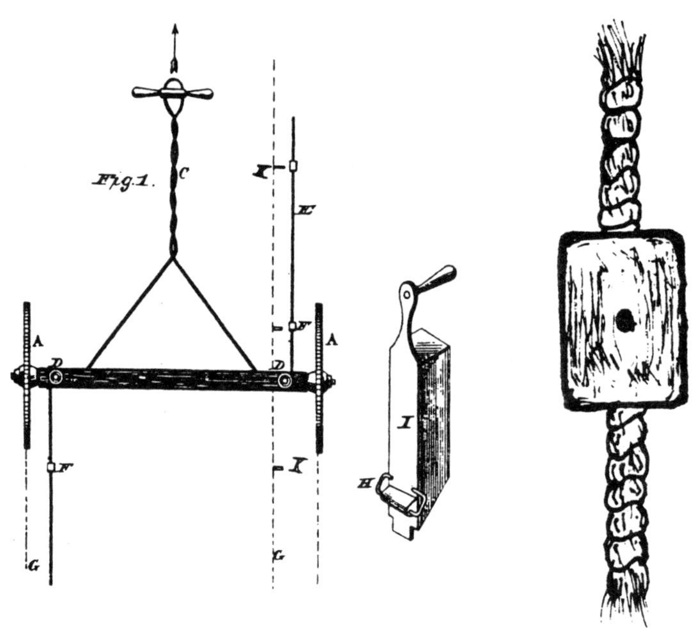

Decatur, Macon County, Illinois

"The wheels of the truck are placed at such a distance apart that the tracks formed in the ground thereby will serve as guides to the operator who presses the point of the planter into the ground depositing the seed near or in the track of the wheel, and at points indicated by the knots or other marks on the cord, thus planting the corn in rows not only parallel with, but also at right angles to the line of the cord or marker and to the path of the operator, producing what is termed 'check-rowing'."   The marks or knots, though claimed, were not described.

November 6, 1877   Chain   196,849

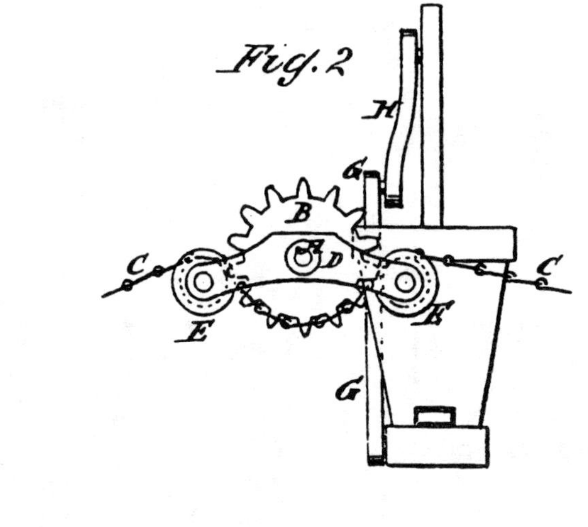

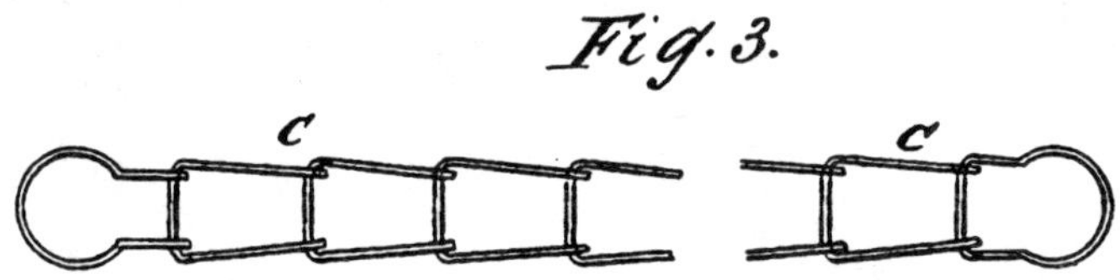

Whitewater, Walworth County, Wisconsin

"To one end of the shaft $\mathcal{A}$ is attached a chain-wheel $\mathcal{B}$, the teeth of which mesh into the links of a chain, C, extended across the field and secured at its ends by stakes, anchors, or other suitable means."

The chain drives the checking attachment. No detail of the chain construction was offered other than the Fig. 3.

JOSEPH HARVEY

November 13, 1877   Cylinders   197,125

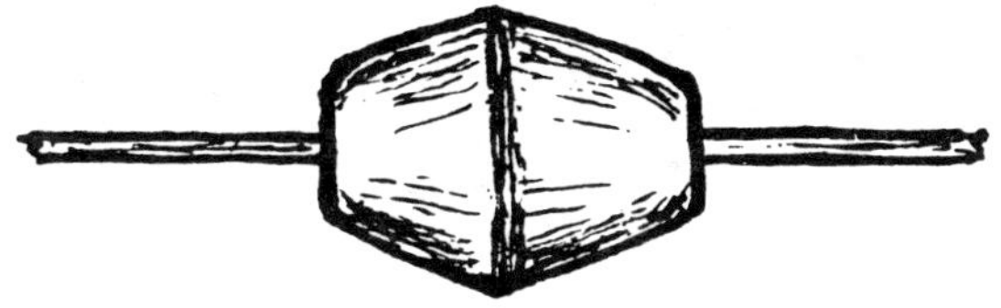

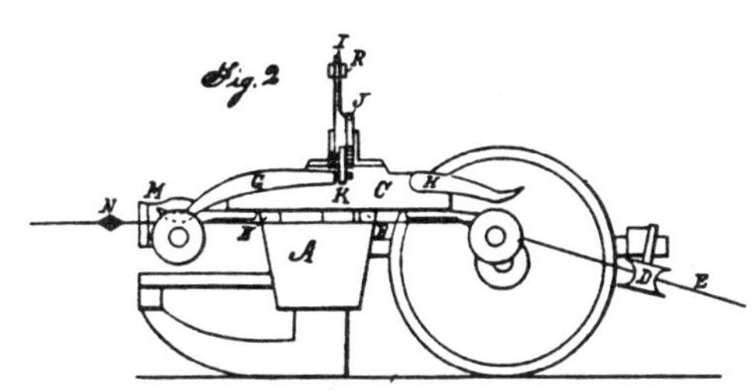

Aledo, Mercer County, Illinois

"The wire is provided with cylinders of wood or iron N, tapered at either end and secured on the wire at a distance equal to twice the distance between centers of the sheave-rollers."

November 20, 1877   Jointed Knots   197,271

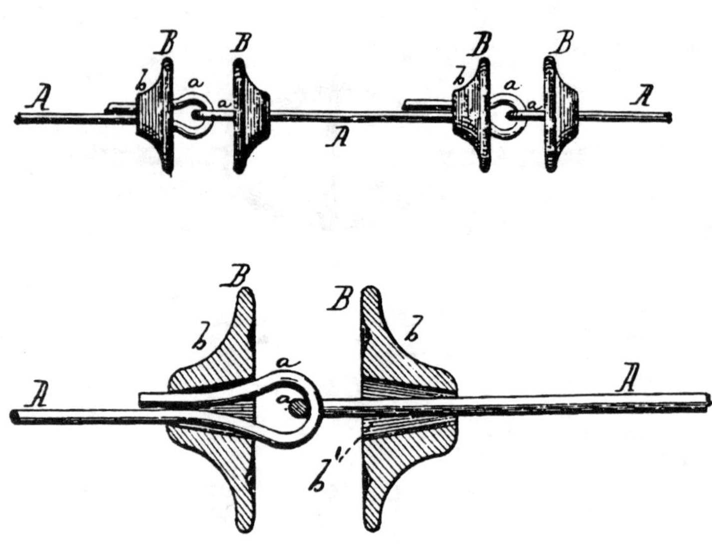

Decatur, Macon County, Illinois

"This invention relates to the manner of uniting the sections of wire by the aid of the buttons forming the knots or stops in said wire and consists in providing the ends of the lengths of wire with loops for joining the sections, and with collars or buttons adapted to be slipped over said looped ends, for preventing them from being drawn out and detached by the tension and strain on the wire."

Haworth goes on to explain that this design may use a smaller gauge wire and that if a wire is broken it is easy to repair without disturbing the knot position.

August 6, 1878　　　　Stops　　　　206,702

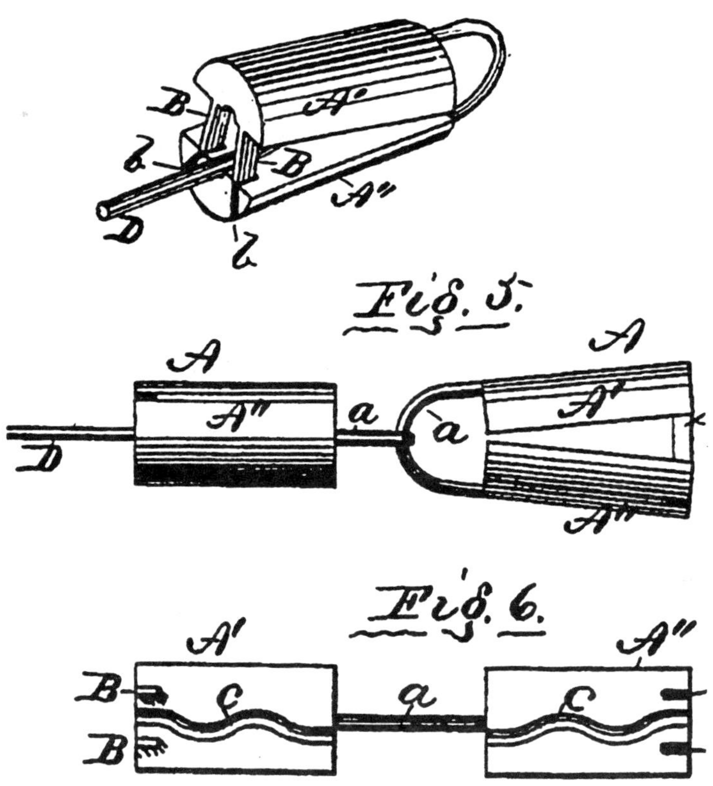

Decatur, Macon County, Illinois

"The semi-cylinders $A'A''$ may be cast or otherwise formed upon the wire a while it is straight, as shown at Fig. 6, or while it is partly bent; or the wire a may be attached to the semi-cylinders $A'A''$ by any desired method; or the semi-cylinders and wire may be made integral of malleable iron.

September 3, 1878   Stop   207,683

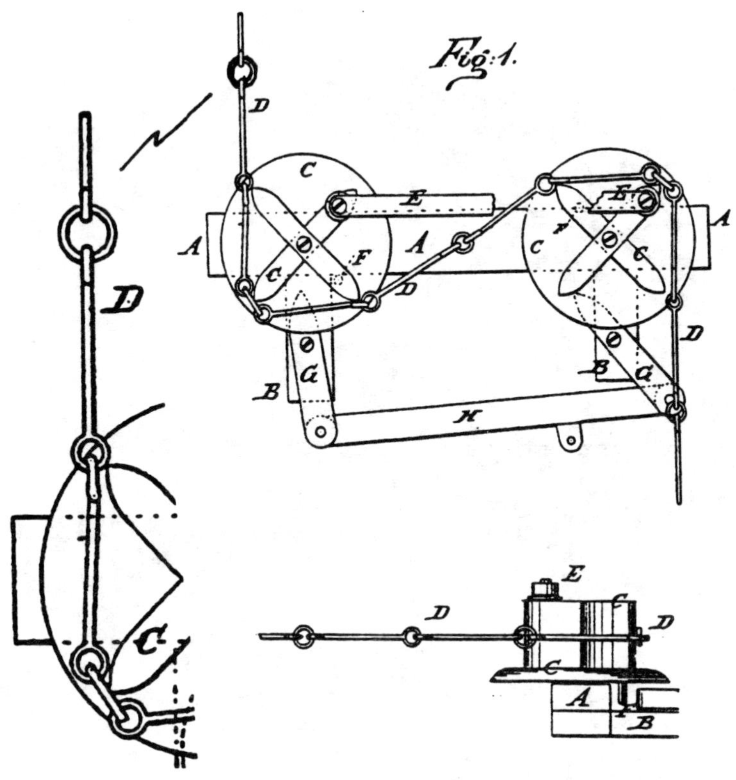

Kent, Union County, Iowa

" . . . joints of the wire chain D, and the links of which are made two feet, more or less, in length equal to the distance between the ends of the arms of reel C. The chain should have a number of swivels formed in it to prevent it from twisting or kinking."

Fig. 1 shows a swivel in every other link.

October 8, 1878   Swivel Knot   208,814

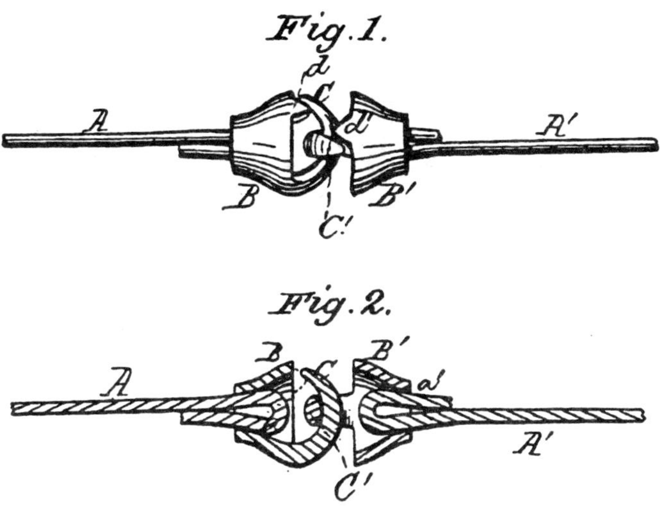

Decatur, Macon County, Illinois

This patent is an improvement over 159,177. The wire swivels within the bell-shaped knot coupling half. The shape allows for easier passage of the knot through pulleys and less strain on the wire.

October 8, 1878   Removable Knot   208,815

Fig. 1.

Fig. 2.

Decatur, Macon County, Illinois

This patent is for a knot that is applied on a continuous cord. Two pieces with hooks are placed over the cord at any spacing desired. They may be removed and replaced for changes in spacing.

CHRISTOPHER G. CROSS

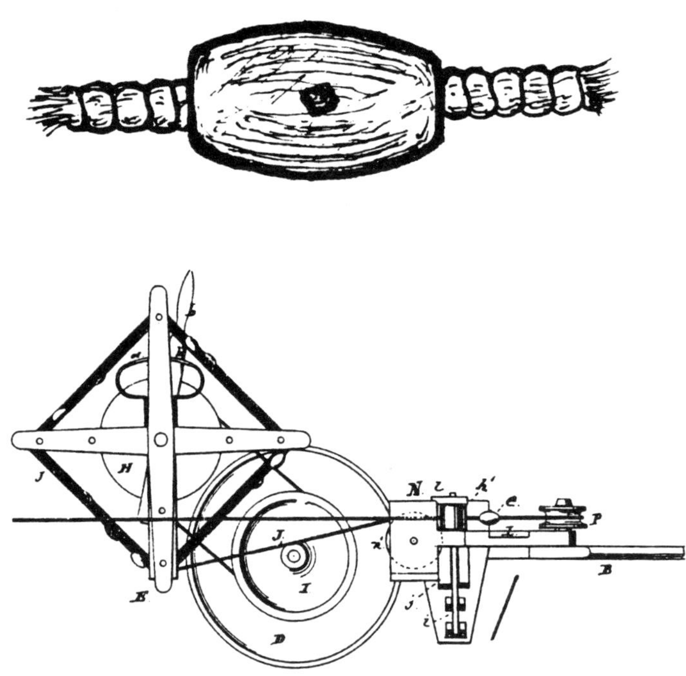

Chicago, Cook County, Illinois

" . . . as the machine passes along the knots or balls e will pass, hence the distance between the balls is double that between the plants."

No description is offered for the ball. It looks to be a wooden, oblong knot on a cord.

GARITT J. HYER

February 18, 1879   Marks   212,469

Clinton, DeWitt County, Illinois

"*A* represents <u>the rope, which is not knotted, but marked at even distances</u> and stretched across the field. The long rope is used to drive the pulley, has marks upon it for its entire length and the distance between the marks is the same as between the rows. The object of these marks is to determine the dropping point when the check-row has to be locked when crossing the field. As the field has not been marked, there is nothing to indicate where the rows are located, and it is obvious that the driver would lose his calculations without some marks upon the long rope to enable him to decide the exact point in a true line with the cross-rows, where to begin again to drop the corn. It is intended that the marks on the field-rope will come opposite a certain point on the pulley at every revolution."

It is not clear in the figure what is used to mark the cord. <u>The mark had to be obvious and may have been colored cord, cloth, or even baling wire</u>; something that would not jam in the pulleys. The figure shows something physically wrapped around the cord.

Hyer added an interesting observation:

"In all prairie country experience is beginning to show that more corn is raised to the acre, and at less cultivation, by drilling the corn in rows running only in one direction instead of planting it in rows at right angles by means of check-row."

GARITT J. HYER

February 18, 1879   Marks   212,469

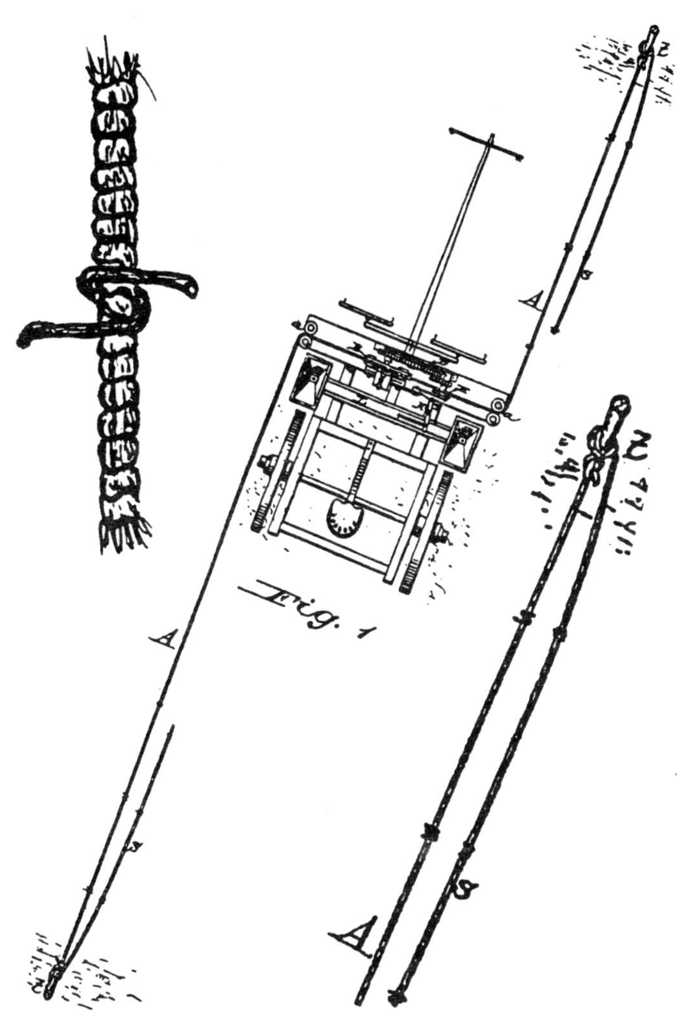

" $\mathcal{A}$ represents <u>the rope, which is not knotted, but marked at</u> <u>even distances and stretched</u> across the field."

February 25, 1879   Knots or Stops   212,553

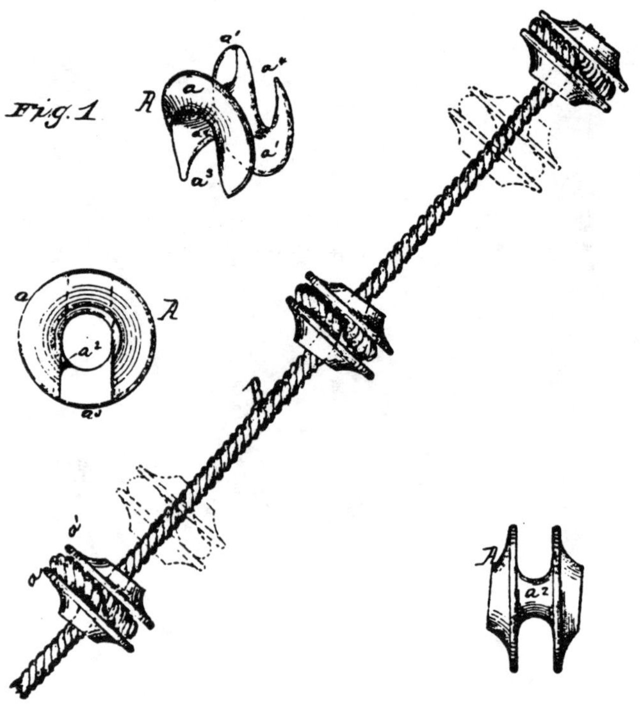

Decatur, Macon County, Illinois

"My invention relates to a novel construction of the knots or stops for check-row cords, permitting the knots or stops to be readily applied to and adjusted upon or removed from and replaced on the cord without necessitating the bending or clamping of same thereon. This may be done by hand without tools. This makes a great savings in breakage and also facilitates the changing of the knot for planting different widths of rows or if the rope should become stretched."

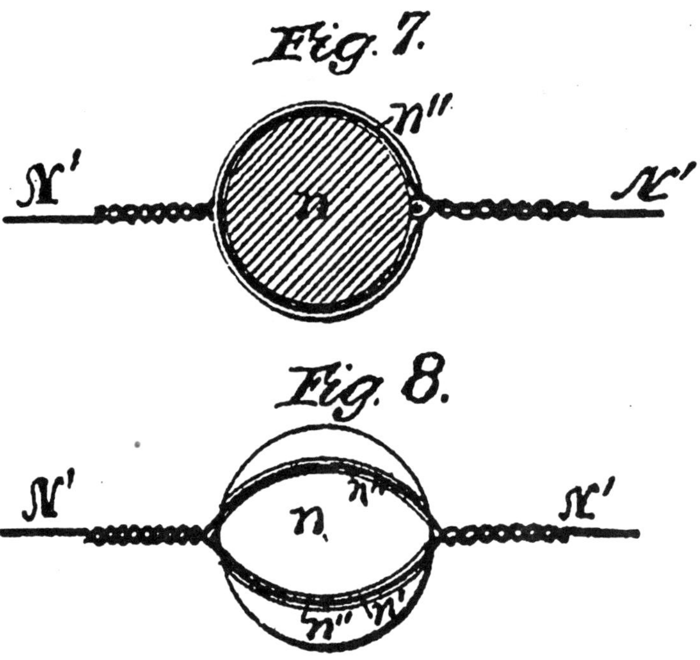

Galesburg, Knox County, Illinois

"A wire check-row chain, formed of short sections, with spherical knots secured between the sections in interlocking eyes formed on the sections, ..."

Aledo, Mercer County, Illinois

"The planter is then drawn along the cord or wire, and the first ball or knot striking against the lug or projection e will swing or vibrate the cam far enough to allow the next ball or knot to pass on the opposite side of the cam to engage with the lug e' on that side."

As can be seen in Fig. 1, there are two knots close together for each drop operation. The purpose of this is so when the planter is at the end of the row and turned around, the knot position will be in alignment without the usual adjusting to the drop mechanism.

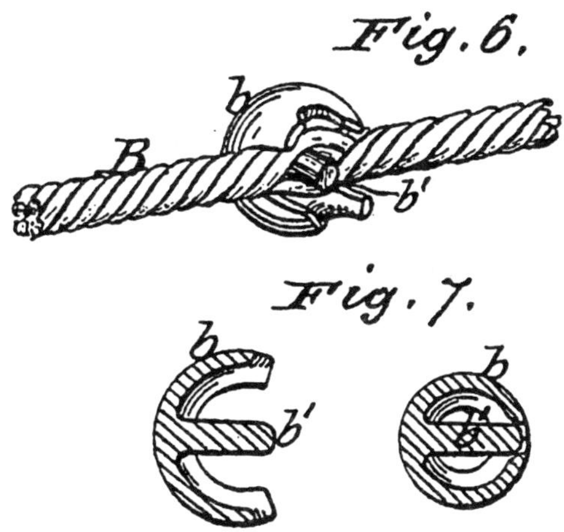

Decatur, Macon County, Illinois

"It consists, also, of <u>artificial knots</u> to be attached to the check rope or cord, said knots being made of malleable metal, first in the form of a crescent or half-ring hollowed in its interior and provided with a radial pin to pass between the strands of the rope and second closed as a ring and compressed around the cord."

September 9, 1879   Loops   219,361

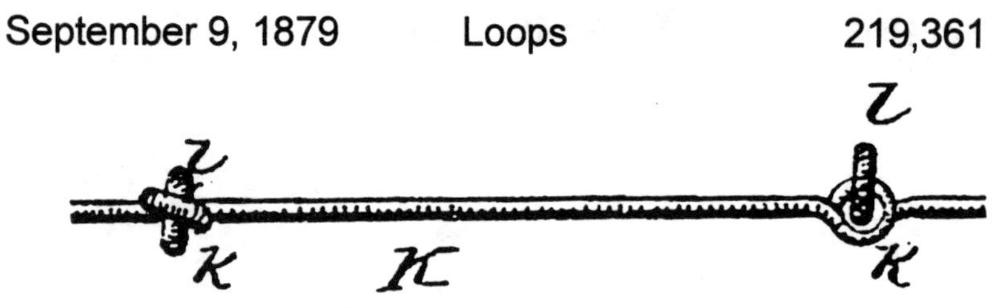

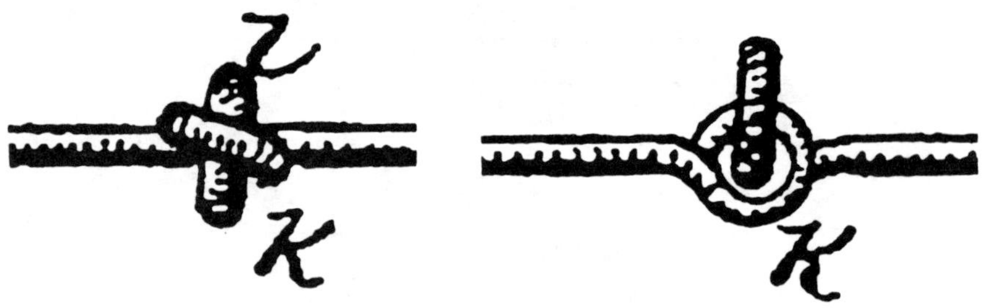

Lisbon, Kendall County, Illinois

"I construct the wire K, by forming therein at desired intervals the loops k, into which are inserted bent wires l, thus forming on a continuous wire loops which cannot slip,.. ... Instead of bent wire l, metallic spurs may be molded on the loops, k. The advantage of a continuous wire is that having no joints, there is no possibility of entanglement, and a consequent change of lengths. I prefer wire to rope for the purpose intended, for the reason the latter is liable to shrink or stretch, particularly if it becomes damp or wet."

Two styles of knot could exist per the above description; one with the bent wire in the loop, and the other with spurs molded on the loop.

MOSES J. BARRON

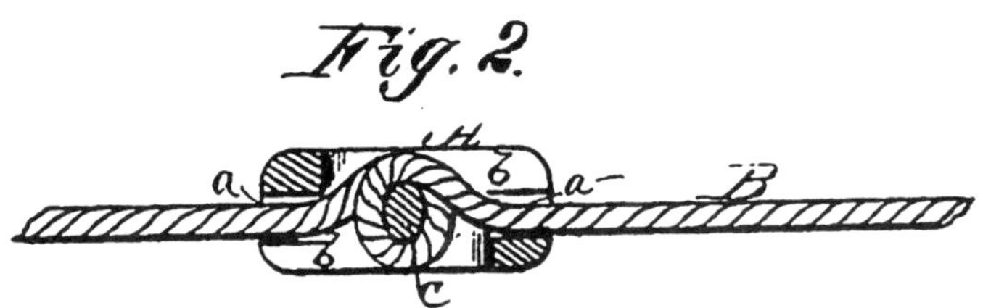

Dawson, Sangamon County, Illinois

"The object of my invention is to provide a cheap and easily attached stop-knot for check-row cords; and it consists in a small piece of malleable metal having a central opening in each end, with a slot through the side, and a central post, around which the cord is would, ..."

October 28, 1879   Stops   221,089

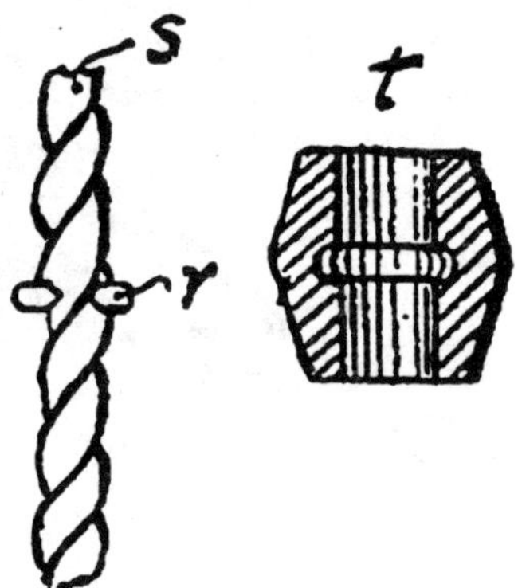

Kenney, Macon County, Illinois

"The stop, Fig. 6, consists in a wire blank r closed in the cord s, and a metal stop t provided with a groove adapted to fit around the wire closed around the same."

"The wire r is bent so that the points are formed which are pressed into the cord. The metal stop is bent around the edge of the wire so as to secure the same."

This patent was re-issued under RE-10,219 on October 24, 1882. However, the stop design was not effected.

December 2, 1879   Tappet   222,278

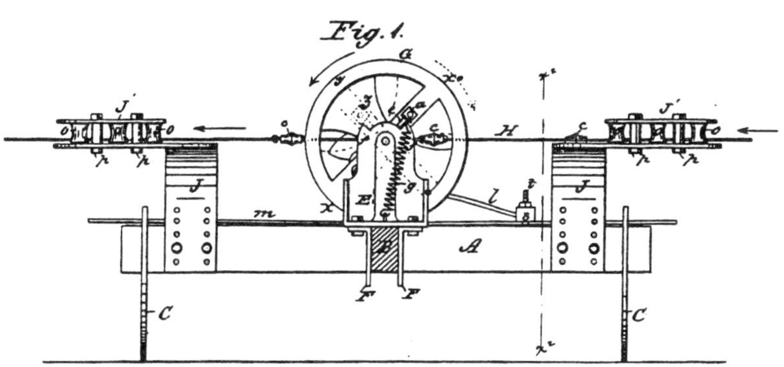

Fairmount, Vermillion County, Illinois

"In the operation of my device I prefer to use a metal chain, $\mathcal{H}$ composed of link-rods having tappets or lugs for the reason that a rope will at times become contracted by dampness and thrown out of check."

No description is given for the tappet. Refer to the author's enlarged view.

December 30, 1879   Knot   223,190

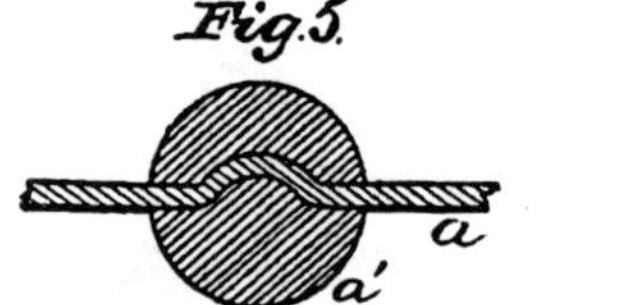

Aledo, Mercer County, Illinois

"For each knot a kink or bend is first made in the wire at the desired point. The kink or bend is then firmly and immovably embedded in any known and suitable way in a ball of metal or other suitable material. If desired the knots may be formed by casting the metal around the kinks or bends."

March 2, 1880   Knot   225,165

Rushville, Schuyler County, Illinois

"The knot is produced by twisting detached pieces of wire around the cord or wire at equal distances apart and then again twisting the cord or wire once around one of these per Fig. 5c. A double knot is formed using two pieces of wire per Fig. 5b."

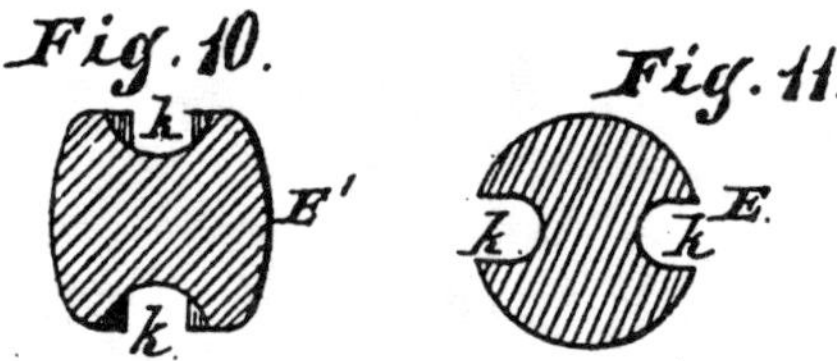

Moline, Rock Island County, Illinois

" *E* is the rope or cable, having knots for operating the lever. The knot is made of metal or other suitable material and provided with a groove, *k*, extending entirely around their periphery, which is of sufficient depth to receive the strands of the rope. The knots are secured to the rope by separating the strands of the rope and inserting the knot until the strands enter the groove, where it can be secured by means of a fastening cord or wire. By this means, no slipping can take place."

April 20, 1880   Knot   226,700

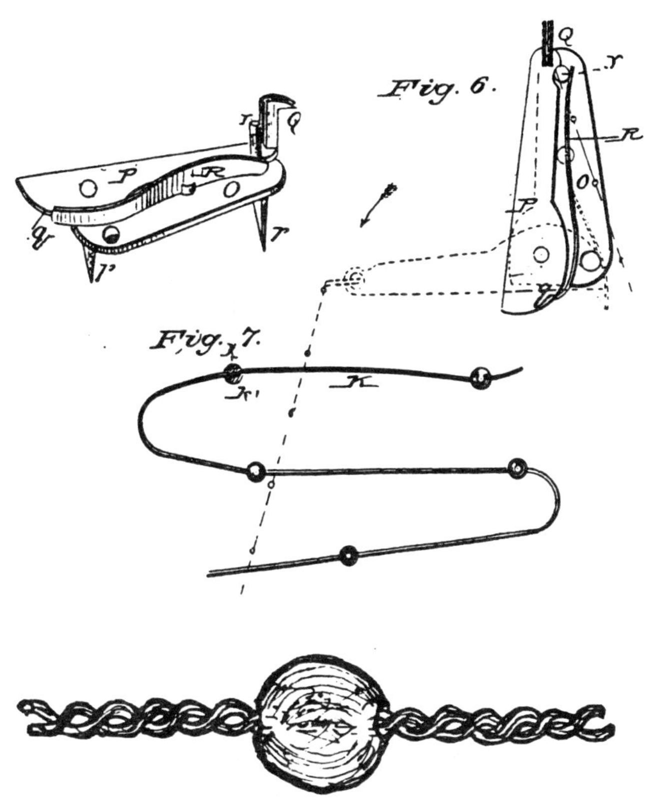

Rushville, Schuyler County, Illinois

"The third part of our invention relates to the operating rope or cord, which <u>we make of continuous lengths of twisted wire</u>, made with equidistant kinks k, around which we cast metal balls k' . . . ."

JOHN W. HUDSON

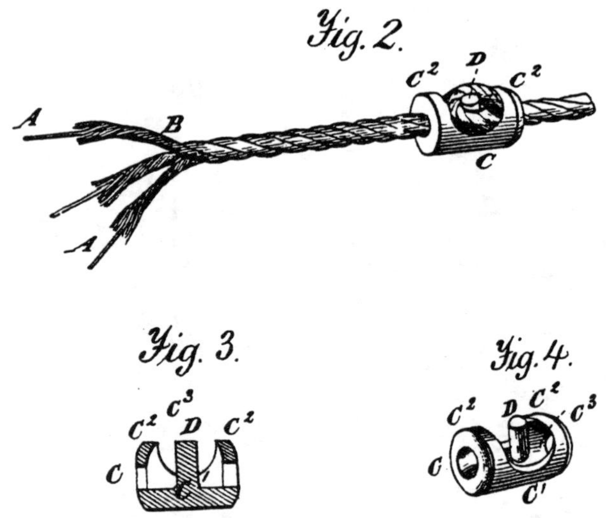

Wellington, Iroquois County, Illinois

This patent includes both the button and the rope with a wire core.

"The button, being made of metal, in one piece, having a central stud, around which the rope is locked in such a manner as to secure it in any desired position by the bite of the rope upon the stud. The button has a cylindrical body and constructed so it can be removed or detached at pleasure."

Note that the use of wire core cords is the same as patent 151,590. It is curious that the patent examiner overlooked this. Hudson should not have been allowed Claim 2 for the cord portion.

JOHN BRICKETT

June 22, 1880   Knot   229,028

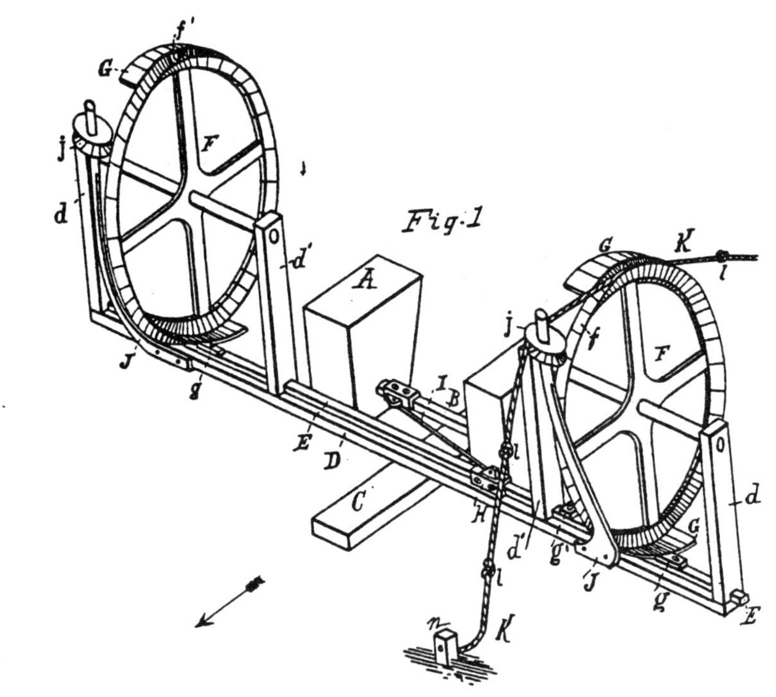

Danforth, Iroquois County, Illinois

"The object of my invention is to produce a simple and effective attachment of few parts, requiring no abrupt curves or bends in the lines used for operating it, thereby admitting the use of a stiff and durable wire cable or cord instead of a pliable rope, which is soon worn out by friction or of linked rod, unwieldy in their operation."

JAMES S. WOOD AND O. WILLIS VAN OSDEL

July 27, 1880   Band or Strap   230,456

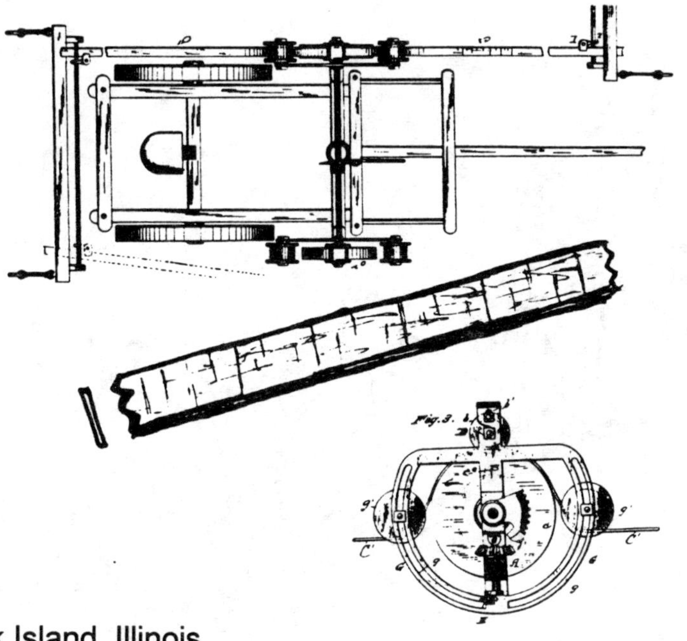

Rock Island, Illinois

This is a metal ribbon friction drive check-row planter. It represented a unique idea.

"The operation is as follows: The check-row band being passed over one of the pulleys, and stretched across the field will communicate motion to the seed slide. When the machine has crossed the field, the check-row band is transferred to the pulley on the other side. "

"We are aware that wire, rope, and chain have been used in check-rower attachments for corn planters but they are objectionable because wire kinks and breaks, the rope stretches and shrinks, while the chain is too heavy."

TYLER C. LORD

September 14, 1880   Joint or Knot   232,137

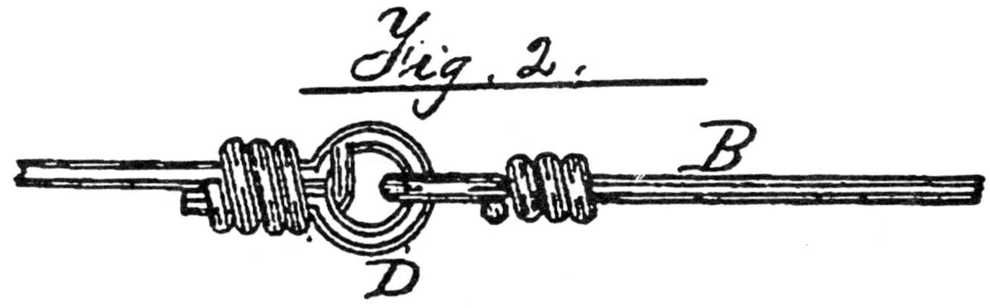

TYPE-1

TYPE-2

Joliet, Will County, Illinois

This patent is for two distinct types of knots.

"The cable joints or knots D are formed as shown in Fig. 2, the short ends running parallel with the main strand, <u>and not wound around it, but held in place by a separate short bit of wire</u>, one end passing through the loop to hold the binding-wire on and in place; or the joint D may be formed by twisting the two ends of the cable wire together, as shown in Fig. 4."

September 14, 1880   Knot   232,151

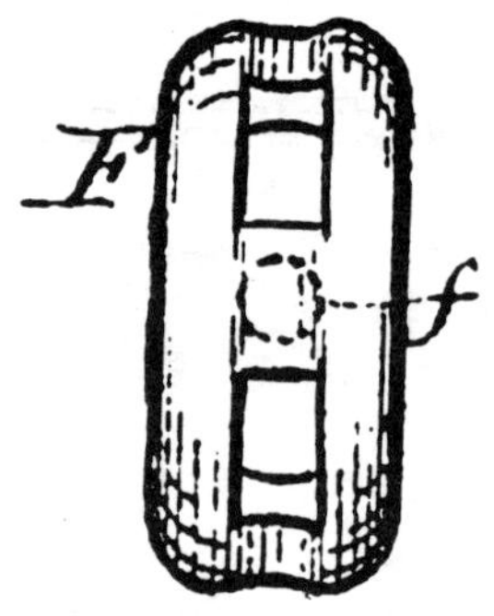

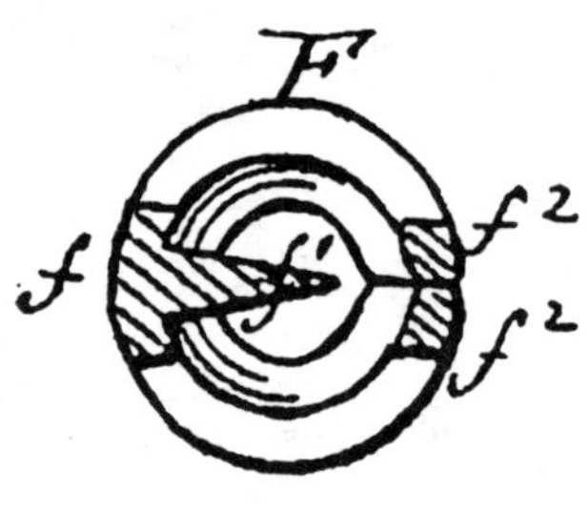

Decatur, Macon County, Illinois

This patent is an improvement over Simpson's earlier patented knot 217,751 on July 22, 1879. Due to the closeness of patents, it is likely that the earlier variety knot was not made in quantity.

September 21, 1880   Stop   232,472

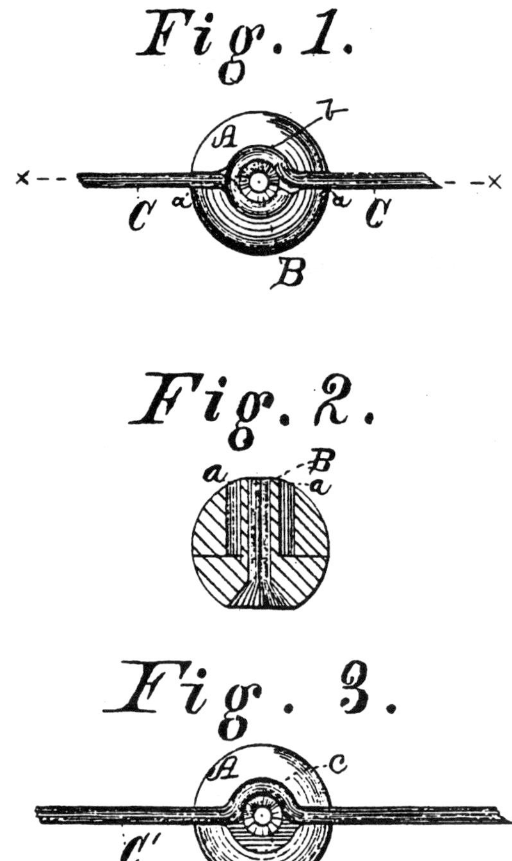

Decatur, Macon County, Illinois

"My invention is an improvement in malleable metal stops; . . . and it consists of a spherical ball provided with suitable channels for securing the wire or wires."

October 26, 1880   Cord-Stop   233,717

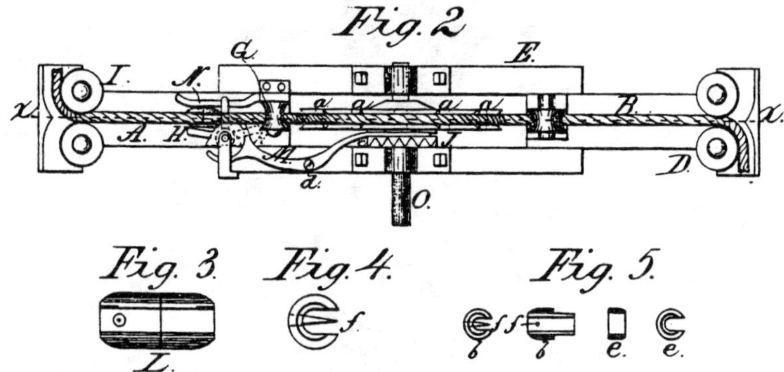

Decatur, Macon County, Illinois

This is an unusual one in that it eliminates the need for "knots or buttons." Instead, a cord drives the mechanism by friction on a pulley. The stop is a removable design that is put on the cord at each end for the purpose of starting and stopping the pulley by triggering the clutch.

Per Fig. 5, b is placed over the cord and the pin f, pierces the cord holding it fast. The small end of b is threaded as the split tube e. Per Fig. 3, the split tube e is then placed over the cord and screwed onto tube b. The split openings of the two parts do not align, thus preventing the cord from coming out.

ANDREW J. GRUSH AND JESSE LOCKHART

November 2, 1880   Stop   233,927

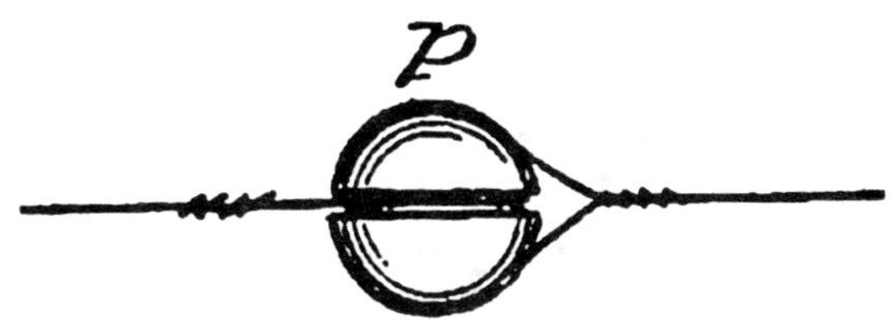

ROUND STYLE

SQUARE STYLE

Macon County, Illinois

" d is the stop on the wire, consisting of a metallic ball having its surface divided by two grooves into four equal parts, and the wire extending around these grooves as shown in Fig. 3, forms a universal joint while the wire lying in the grooves is protected from wear."

The variation sketch is a square style that meets patent specifications.

RICHARD E. CAVINESS AND GEORGE McCORMICK

November 9, 1880   Chain   234,243

Beckwith, Jefferson County, Iowa

"The links of the chain $\mathcal{H}$ are made of a length about equal to the distance apart of the hills, and the ends of the adjacent links are connected by small rings or links."

See Fig. 1, and the author's sketch.

GARITT J. HYER

November 23, 1880   Linked-Chain   234,780

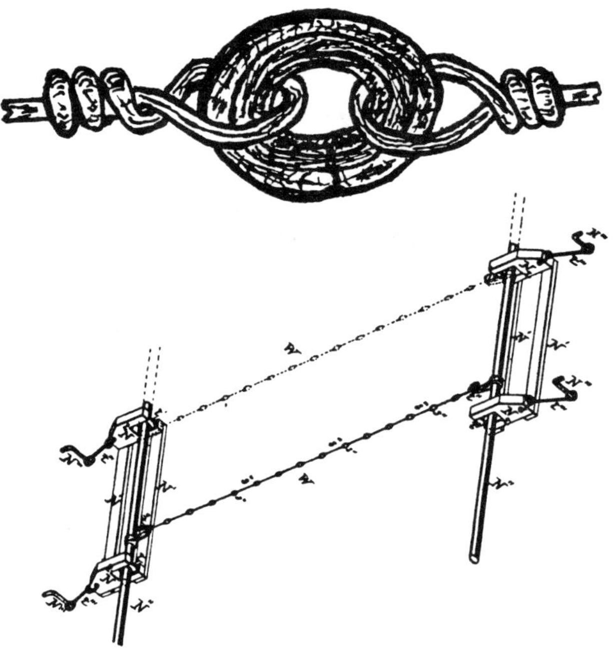

Clinton, DeWitt County, Illinois

"This invention relates to check-row machines employed mainly in connection with what is known as 'horse-planters' or 'horse-power corn planters,' and in which a linked chain, knotted line, or other device is employed . . . . "

"At M is represented a portion of a chain of suitable construction to operate my machine; and it consists of a link s'', made in any suitable manner and of such size as to prevent its passage, and of wire links, s'', looped therein, producing a chain with the enlarged links. . . . "

December 28, 1880   Coupling-Button   236,025

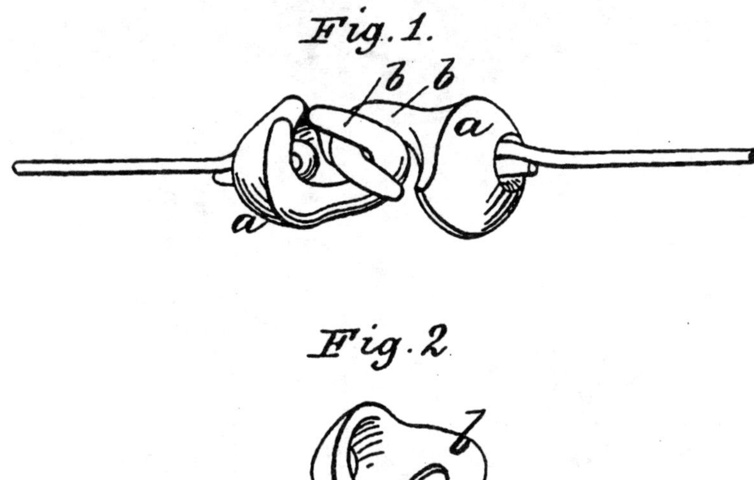

Decatur, Macon County, Illinois

This patent is an improvement over 208,814.

"It has been usual heretofore to cast the parts of the knot for check-row cords or wires with the eyes formed in it and a spur upon it, which, after being annealed or rendered malleable, was bent into the form of an open hook. This process was found to be expensive and tedious, and by making slight modification in the form of the parts I am now able to cast the coupling hooks in the exact form required for use . . . ."

March 1, 1881   Coupling-Stop   238,370

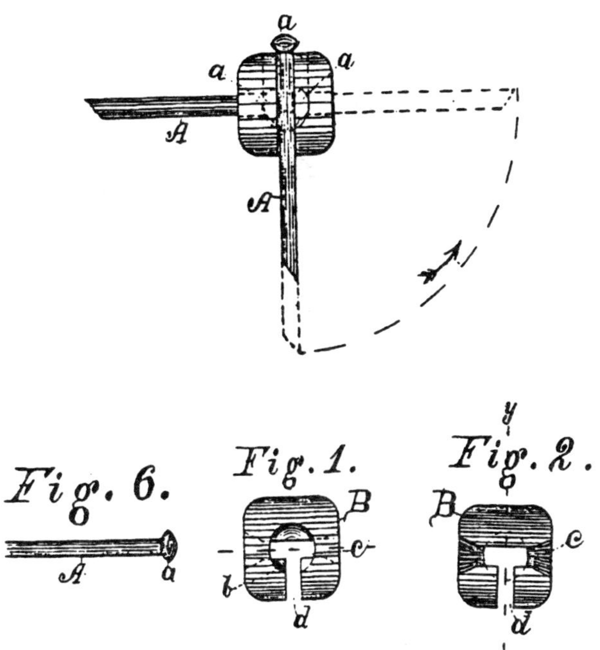

Decatur, Macon County, Illinois

". . ., and it consists, first in the construction of a coupling-stop out of a single piece of metal, provided with a hole of suitable diameter to receive both heads of the wire therein, and slots to allow the placing or removal of either the stop or the solid-headed wire without changing the form of either; and, second in the construction of a solid head out of and on each of the sectional wire, to allow the placing, removal, and swiveling of the head in the stop without changing the form of the head."

July 12, 1881   Knot or Tappet   244,178

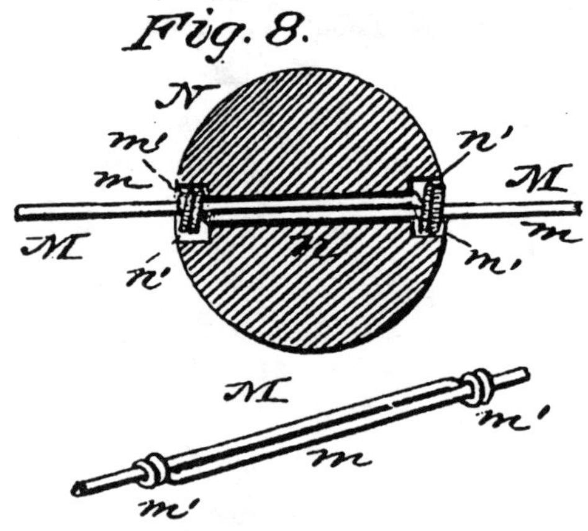

Galesburg, Knox County, Illinois

"M is the check-row chain or wire, formed of sections of m of wire, with a <u>spherical or otherwise suitably-shaped knot or tappet</u>, N secured at the unions of each section m with the adjacent sections, as follows: The tappet N has a central hole, n with enlarged ends n'. The adjacent ends of two sections of wire m are passed through the holes n and each coiled upon the other and the coils m' drawn within the enlarged ends n' of the holes n, as shown in Fig. 8, thus securing the ends of the wire sections to each other and the tappets at the union."

August 23, 1881   Knot   246,273

Galesburg, Knox County, Illinois

"The chain is made in short sections, and a metallic spherical ball, m, held between the adjacent ends of all the sections, by grooving the ball and passing of one section around it and coiling it on the same section so to securely hold the ball. A slight groove m', in the ball then permits the end of the wire of the other section to be interlocked with the section-eye which surrounds the ball, and afterward coiled upon itself to secure it to the other section, as shown in Figs. 11 and 12."

December 13, 1881   Knots   250,749

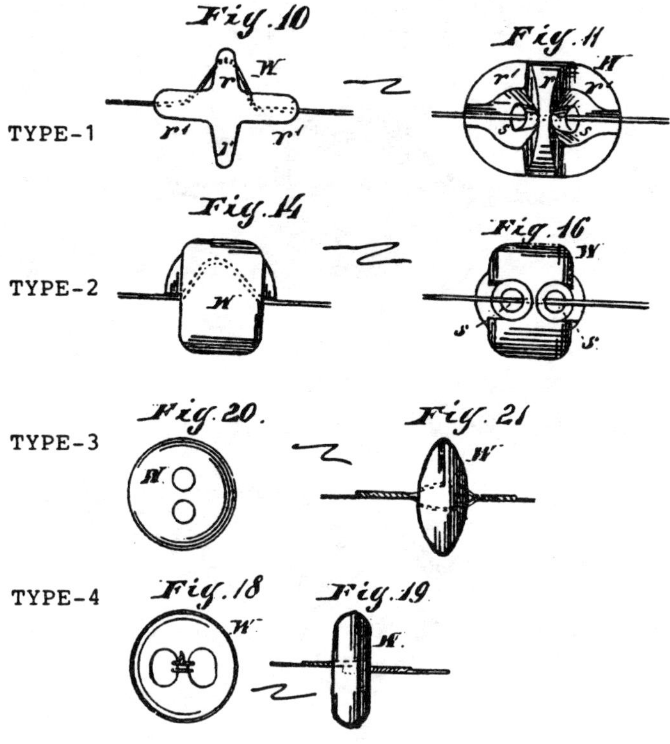

Moline, Rock Island County, Illinois

This patent claims four distinct type knots.

Claim 7: "The check-cord knot herein described, constructed with a central portion cross-piece or partition, having on its opposite sides the openings for the passage of the check-cord, said cross-piece forming a bridge for the cord substantially as and for the purpose set forth."

December 13, 1881   Knots   250,750

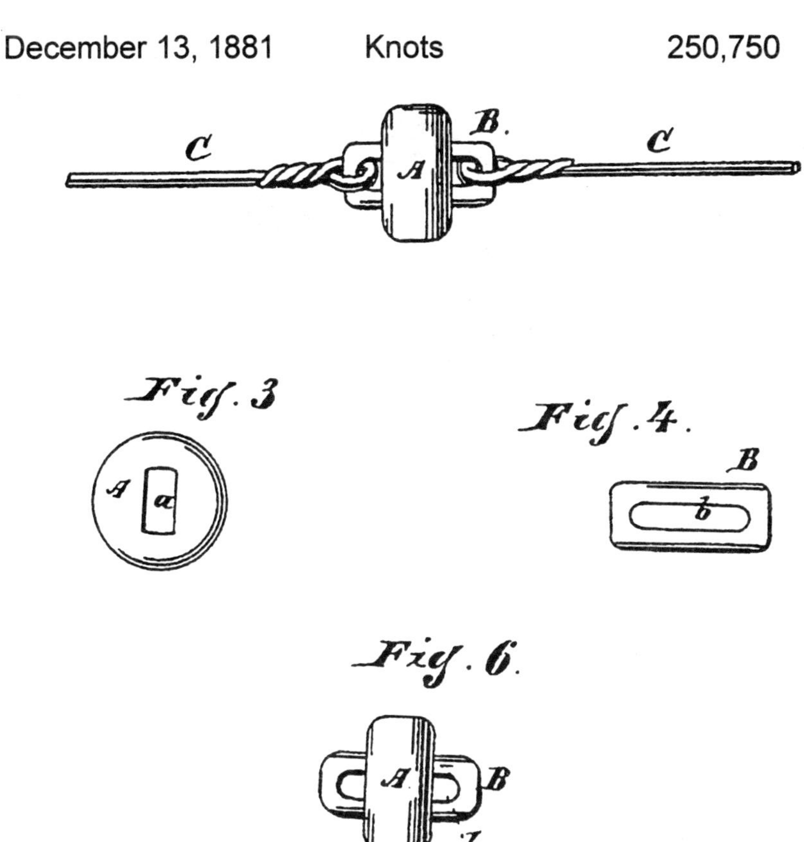

Moline, Rock Island County, Illinois

"In the drawings, A represents the ring or disk portion; B, the link or eye portion; C, the check wire or cord; a, the central opening in the portion A; b, the opening of the link portion B."

Both pieces may be made of cast malleable metal. This knot could be used with both wire or cord.

February 28, 1882   Balls   254,267

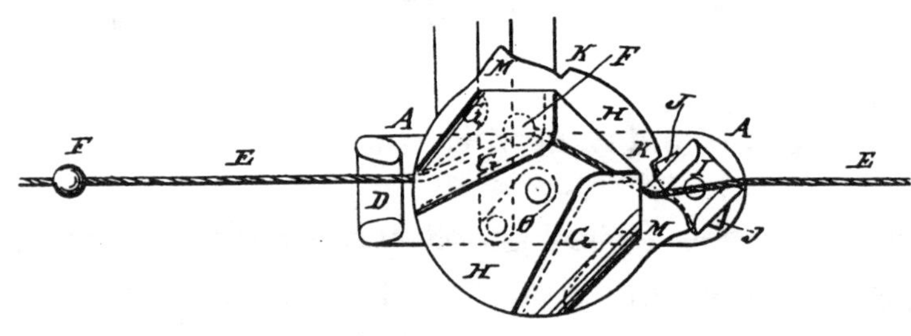

Monmouth, Warren County, Illinois

Reference is made to balls on a cord. No other description was offered.

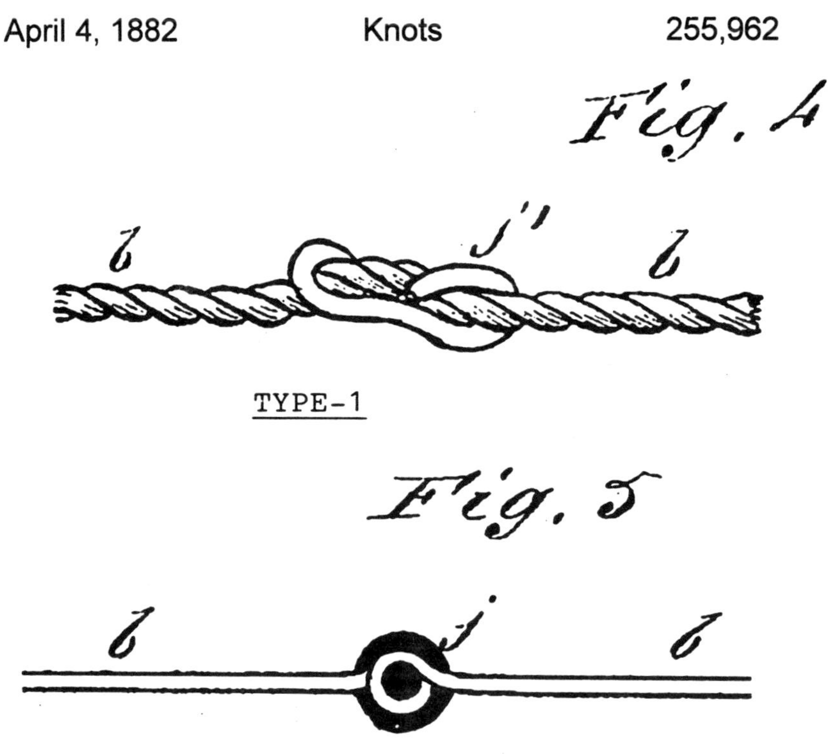

Oskalossa, Jefferson County, Kansas

"In the case of wire ropes, the buttons j, may be of metal balls cast upon bends or coils formed in the said rope, as illustrated in Fig. 5. In the case of ropes made of hemp or other fibrous material, the buttons j, may be formed of pieces of wire the ends of which are bent to pass around the said rope, the planes of the said bends being at right angles with each other, ..."

Two types of knots are possible; Fig. 4 is the wire knot on the rope and Fig. 5 is the cast knot on a wire.

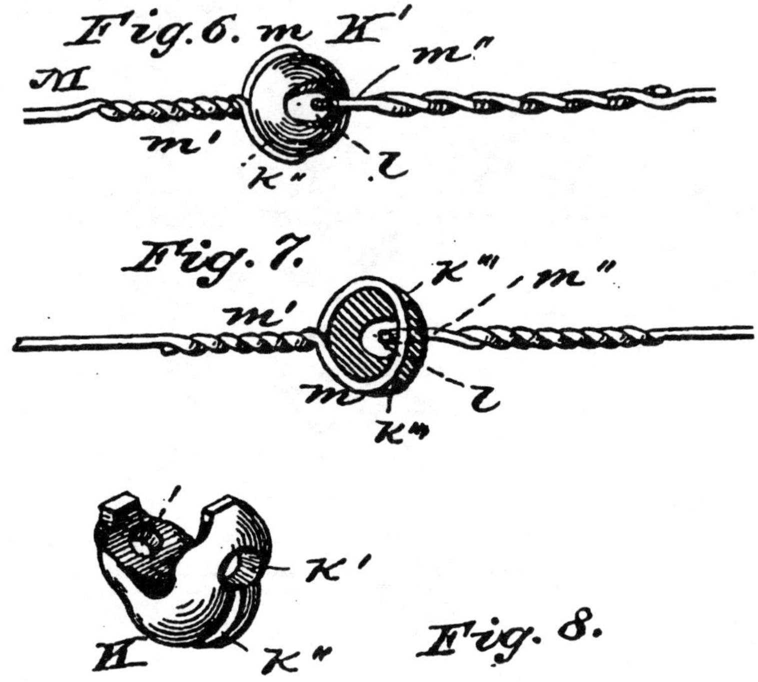

Galesburg, Knox County, Illinois

"Fig. 6 is a plan view of the tappet-wire and one tappet. Fig. 7 is a sectional plan of the wire. Fig. 8 is a perspective of the blank from which the tappet is formed."

The blank is cast malleable iron. Wire m' is wrapped around the blank first, then the blank is pressed closed forming a sphere. Wire m'' is then passed through the hole and twisted upon itself.

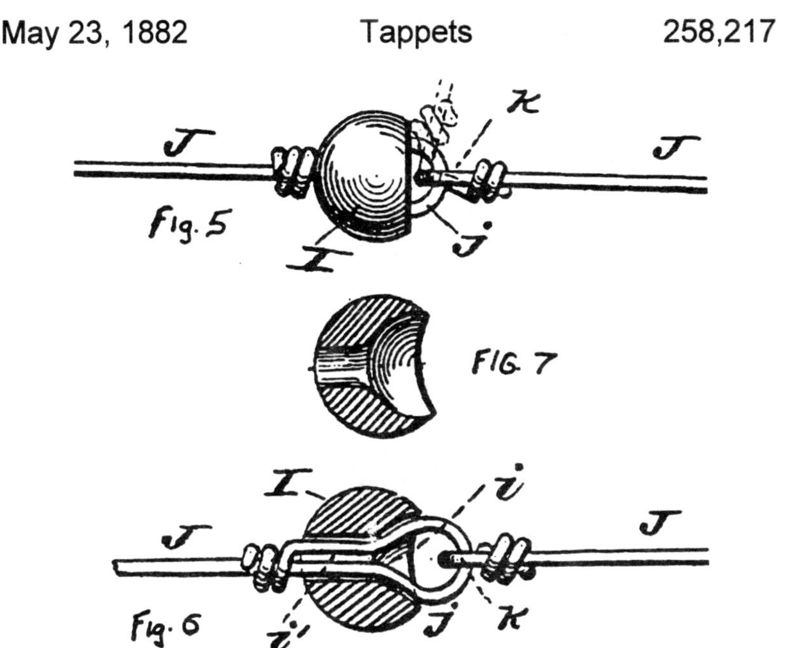

Galesburg, Macon County, Illinois

"The tappets are secured to the tappet-wire as follows: Each spherical tappet I is formed with a cavity, i, in one side, and a hole i, extending from the bottom of said cavity through said tappet. The wire J is formed in sections united at each tappet. The end of one section is bent back to form an eye, j and the ends of this section are then passed through the hole i', so that the eye j rests in the cavity i, as shown, where it is securely held by coiling the end of the wire upon itself, as shown in Fig. 6. An eye k on the end of another section of wire is then interlocked with the eye j and held thereto by coiling its end upon itself, ..."

September 12, 1882   Knot or Stop   264,069

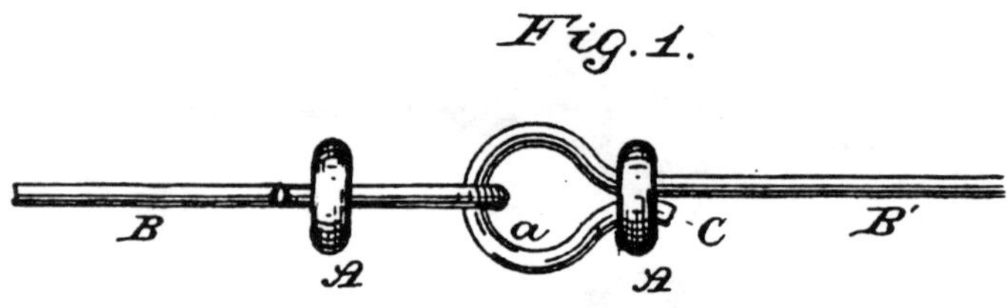

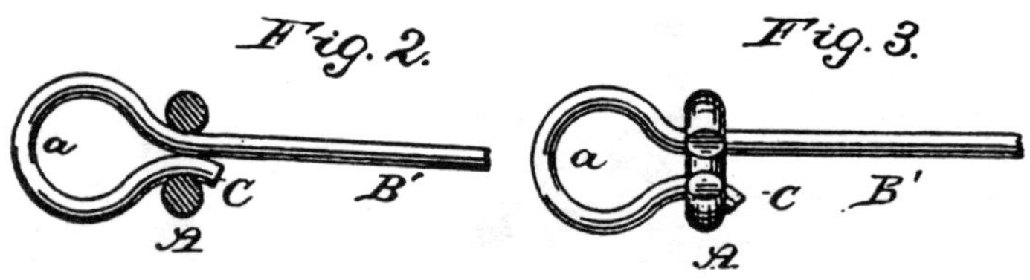

Decatur, Macon County, Illinois

"Fig. 1 shows the two sections of wire BB linked together by loops a and open rings A closed around the returned end C and the main wire B. Fig. 5 is a side view of ring A before being applied, which may be formed of stout wire or cast of malleable iron."

September 12, 1882   Spring Stop   264,313

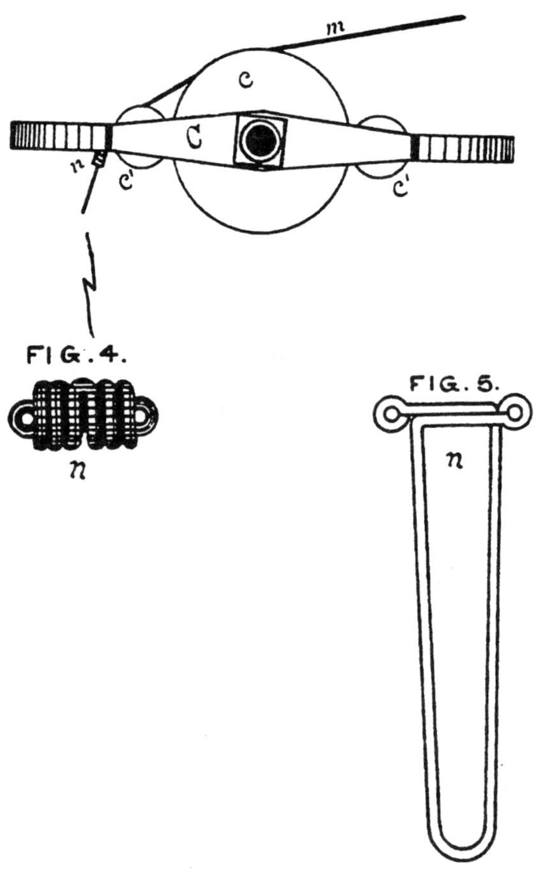

Decatur, Macon County, Illinois

"The stop n operates as a spring to lessen the jar on the tappet."

October 10, 1882   Knot-Stop   265,665

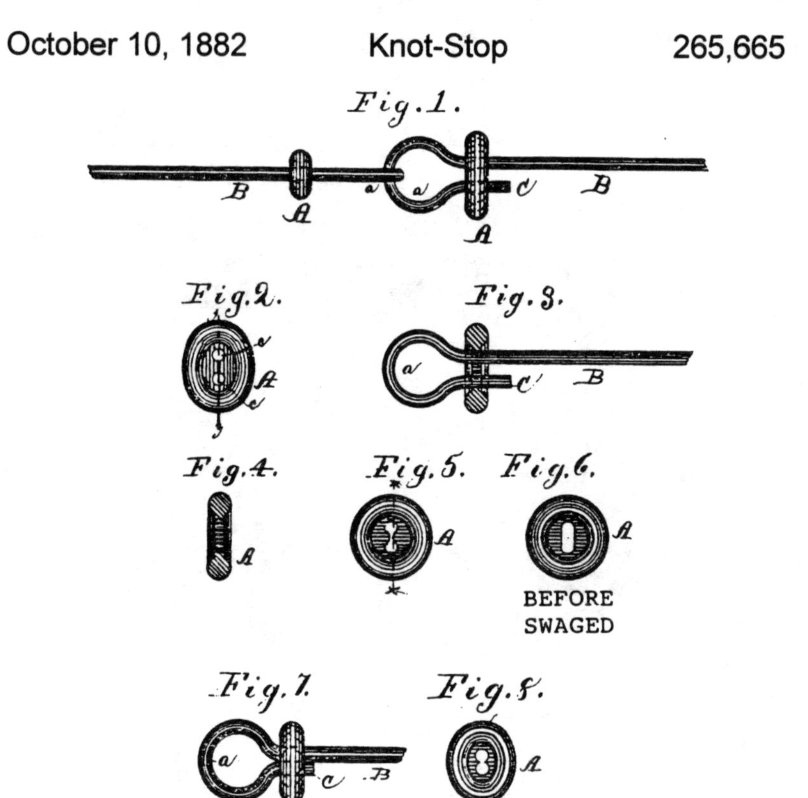

Decatur, Macon County, Illinois

"This improvement consists of forming a solid ring with an oblong hole and a web of thinner metal toward the center for the purpose of enabling it to be more readily slipped over and swaged down onto the two parts of the wire to be secured."

This patent was applied for separately from 264,069, and covers the ring that likely was used on knots in that earlier patent.

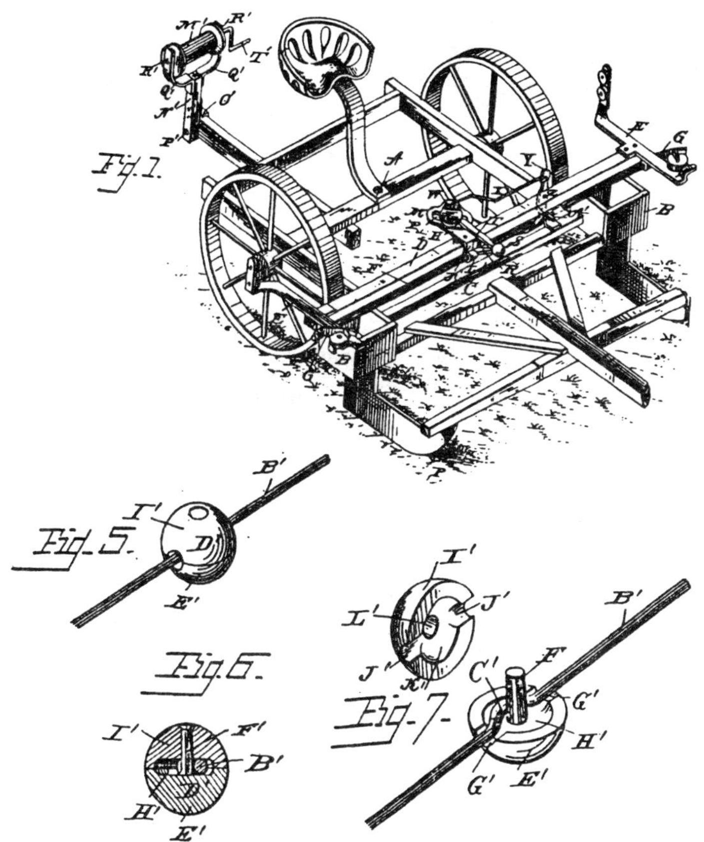

Fairbury, Illinois

This check-row knot is a spherical form made in two pieces per Fig. 7. A kink is made in the wire to rest in the recess of one half with the stud. The other half fits over the stud and completes the knot. The end of the stud is then peened.

October 24, 1882   Auxiliary Knots   266,342

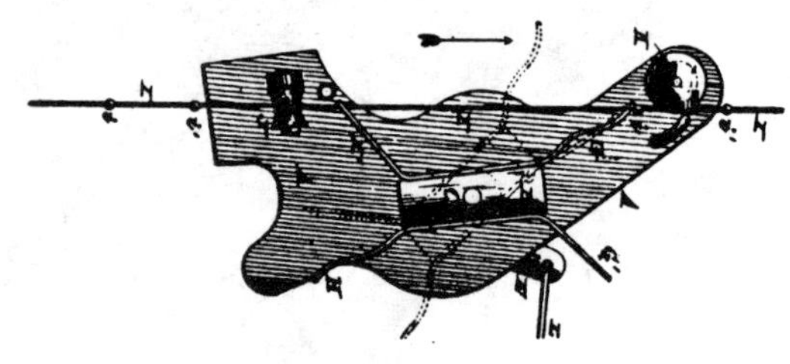

Galesburg, Knox County, Illinois

This patent is for a chain with two knots close together to operate the levers. The knots are described in 267,054. Both patents were applied for on the same date of August 3, 1881.

" . . . and it consists in an improvement in the knotted-wire chain or cord."

"I is the check-row chain, having a series of knots, i, secured thereto at distances from each other equal to the distances between the hills of corn to be planted. "

"i' are a series of auxiliary knots on the chain I ---one to each knot i and at a short distance therefrom."

November 7, 1882   Knot   267,054

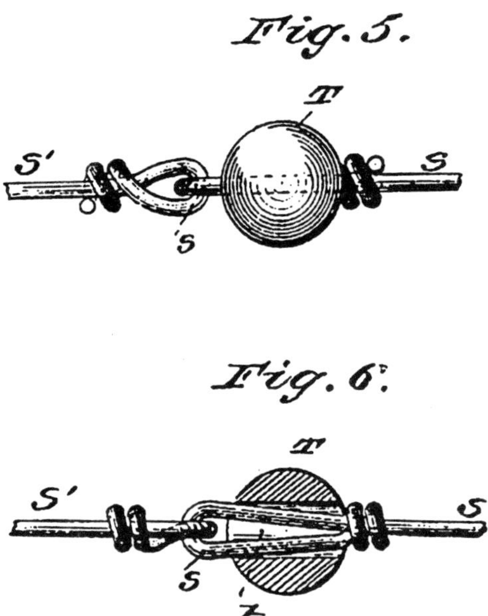

Galesburg, Knox County, Illinois

"The knots T are attached as follows: The end of one section S, of wire is bent back upon itself and passed through a hole, t in the knot T and then its end coiled upon itself as shown at Fig. 5. The end of the other section of wire, S', is then passed through the projecting loop s of the first named section and a slightly enlarged loop s' is formed thereon, which prevents the loop s being withdrawn through the hole in the knot T. The section S' is then secured by coiling its end upon itself as shown at Figs. 5 and 6."

January 2, 1883   Knot   269,940

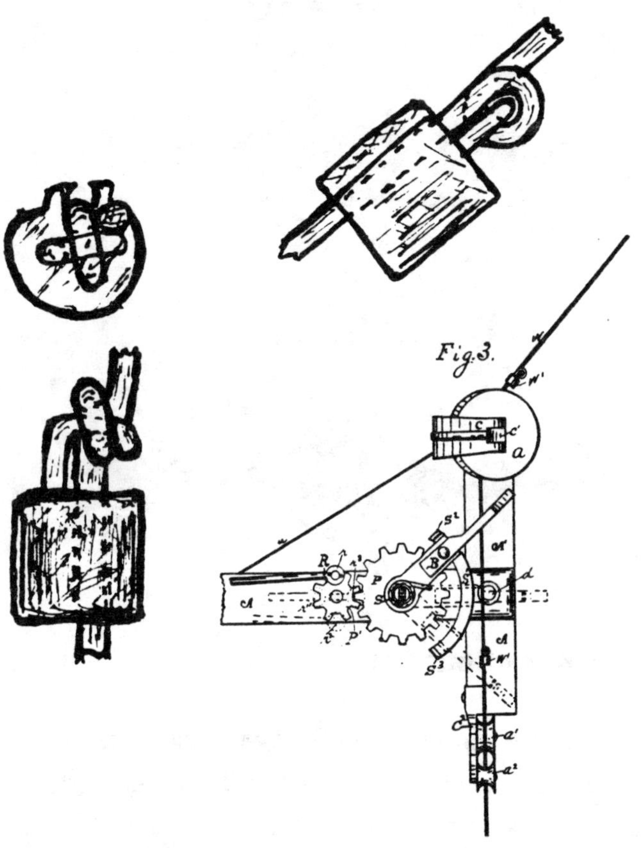

Joliet, Will County, Illinois

The knot shown in Fig. 3, is not described in the patent. It is shown here enlarged as it is a design different from others and may have been manufactured.

RICHARD E. CAVINESS AND GEORGE McCORMICK

January 30, 1883   Trip-Wire   271,419

Beckwith, Jefferson County, Iowa

"Upon the wire E at a distance apart equal to the required distance apart of the hills are formed loops, coils, or eyes F through each of which is passed an open ring G. The rings G are then closed, securely fastening them to the eyes F. The center of the piece of wire of which each ring G is formed may have a groove H formed around it to receive the eye F and prevent said ring from slipping in said eye, even should said ring be forced open by accidental pressure."

February 6, 1883   Coupling   271,788

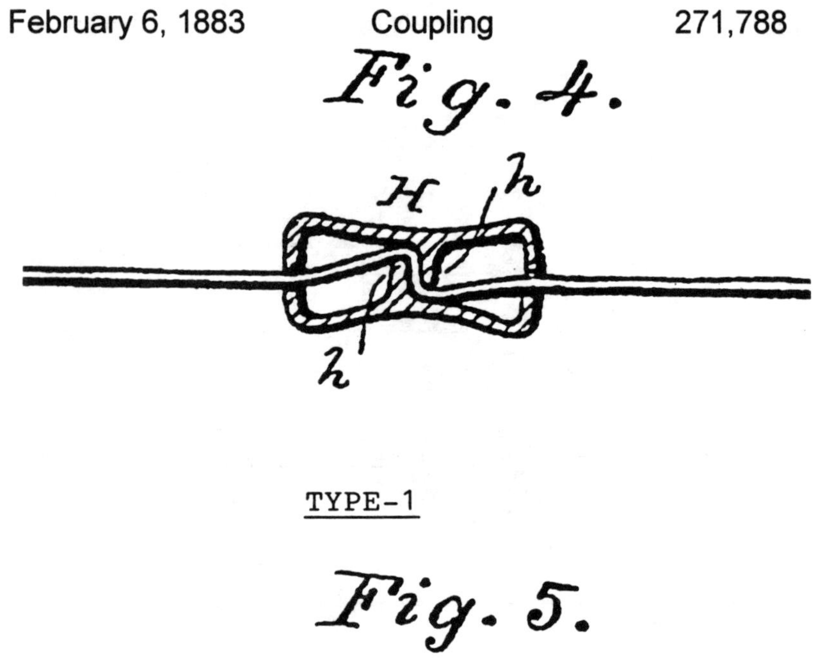

TYPE-1

TYPE-2

Decatur, Macon County, Illinois

This patent is for two distinct type knots. Fig. 4 shows it applied on a continuous wire. Fig. 5 shows it applied to sections.

February 20, 1883   Buttons   272,702

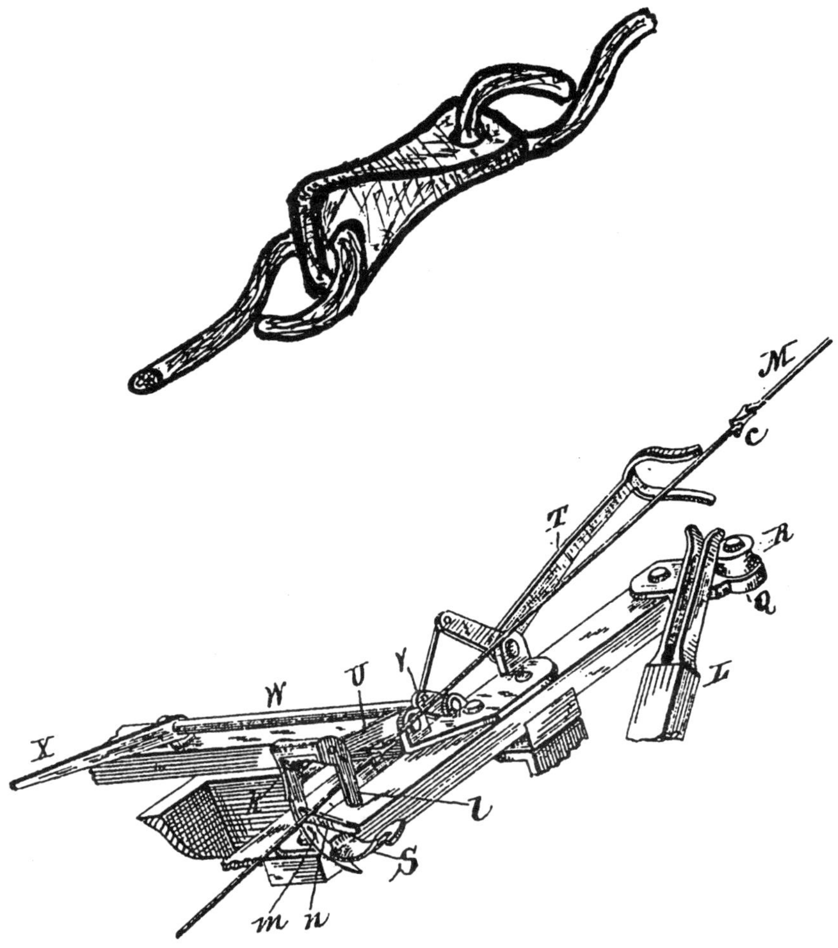

Sterling, Whiteside County, Illinois

The shape of the button suggests it is made as a casting.

February 27, 1883   Check-Row Cord   272,916

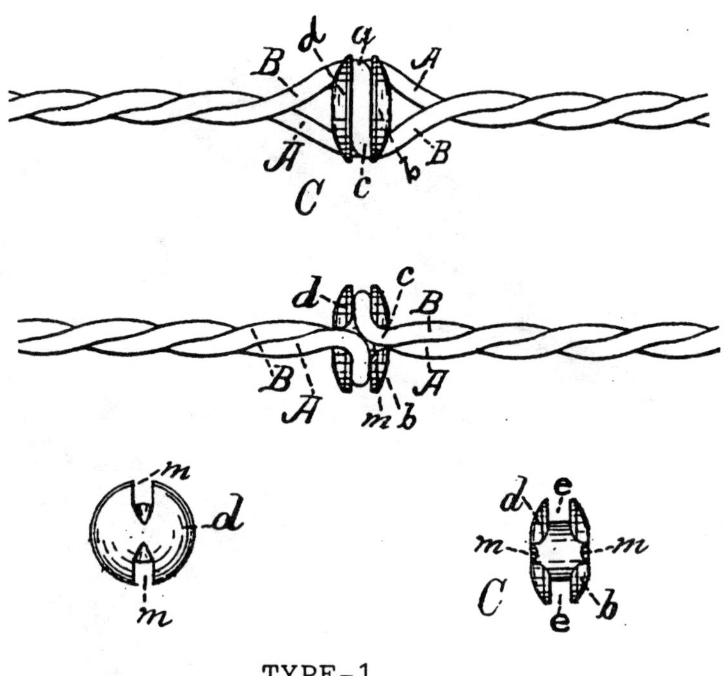

Freeport, Stephenson, County, Illinois

Two type buttons are possible. Type-1 shown is for a two-strand line. Stover says that it is possible to use a three-strand line if the button has three notches for a type-2.

March 20, 1883   Metallic Link/Knot   274,123

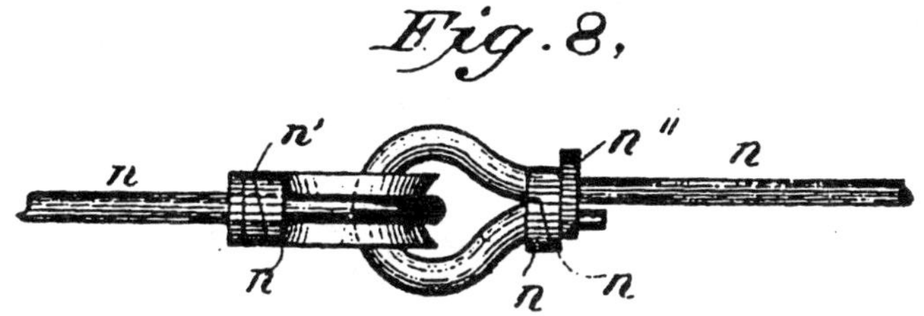

Decatur, Macon County, Illinois

"By reference to Fig. 8, m represents the mechanical link or knot and nn the wires. This loop or knot is made of cast malleable metal and has two short tangs $n'n'$, which, when the wire is bent around the outside of the metallic loop, are bent down over the returned end of the loop, securing them together and to the body of said loop. There is an offset n'' at the end of the body of the loop which is provided with a hole through which the main wire passes before being bent out around said loop. The point of contact is between the two loops and not the wires, which protects the wires from wear."

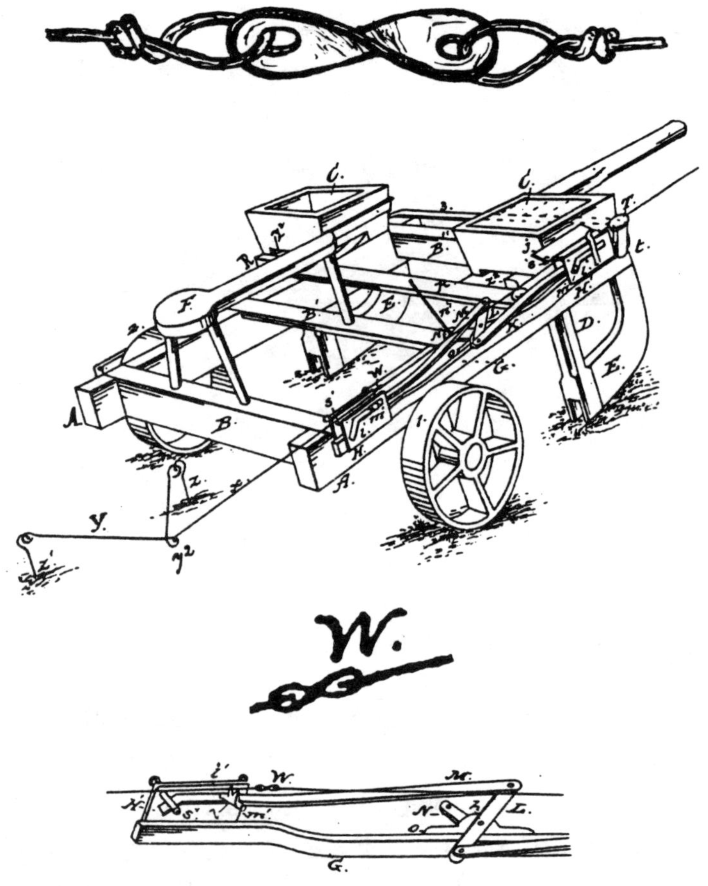

Lytle City, Iowa County, Iowa

"The object of my invention is to lessen the number of parts, ... X the chain or cord is provided with knots W <u>twice the distance from each other of two rows</u>."

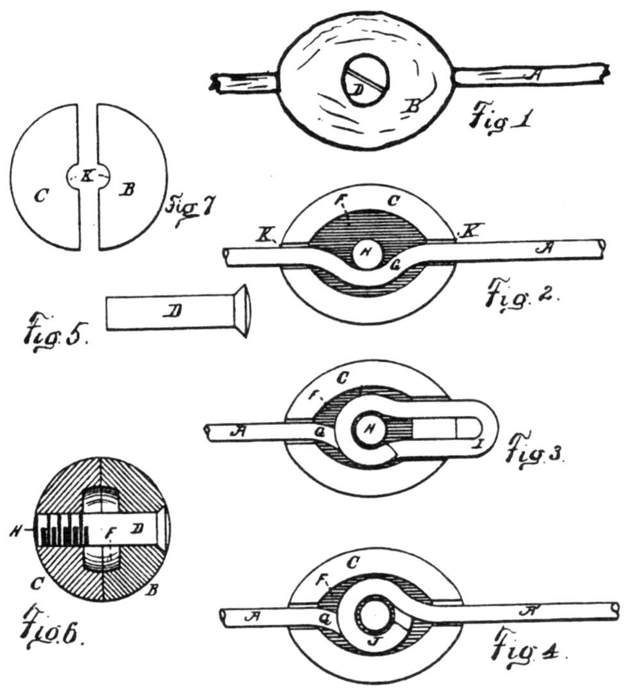

Richmond, Wayne County, Indiana

"This invention relates to the construction of the cord-stop. The stop is composed of two similar halves BC, arranged to be clamped upon the cord A by screw or rivet D that is separable from the halves and passes through them both. Cavities F, of similar form in the inner faces of the halves, allow room for the cord or wire as it passes around the screw or rivet. Openings K at the end of the stop are for the admission of the wire or cord."

May 22, 1883   Stop   277,956

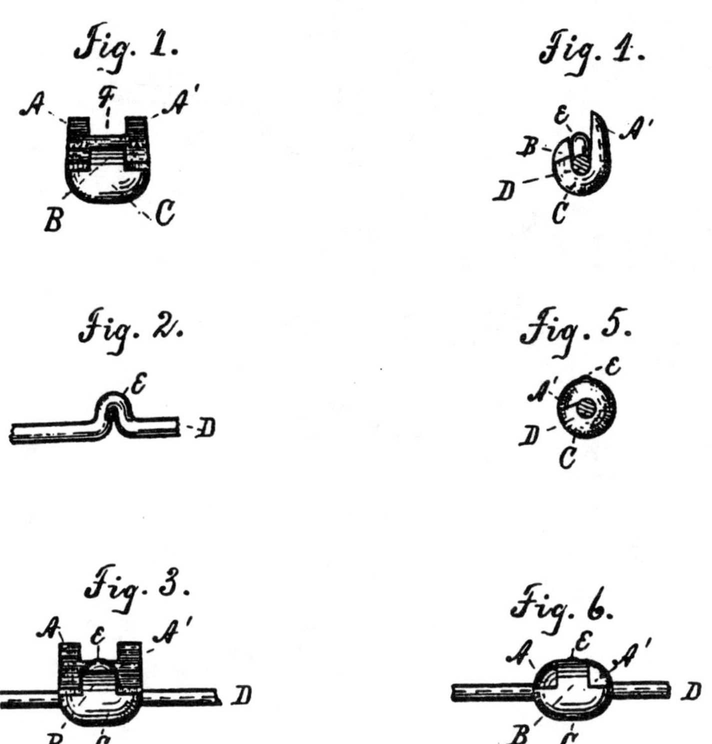

Freeport, Stephenson County, Illinois

"My invention is an improved check-row wire provided at regular intervals with stops of malleable metal, the form of the wire and of the stops, as well as the method combined as per Figs. 1 through 6."

August 7, 1883   Knot   282,532

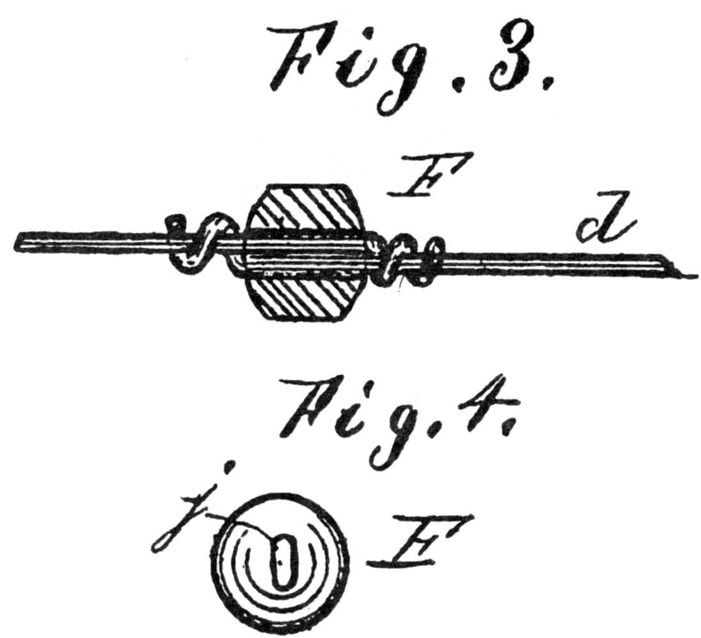

Decatur, Macon County, Illinois

"The knot or stop F is made of malleable metal, with an oblong hole j a suitable size to receive two sections of wire and is secured thereto by simply twisting each protruding end around the main wire as shown.   This forms an enlargement on each side of the knot that cannot be drawn through the oblong hole."

September 4, 1883   Stop/Knob   284,240

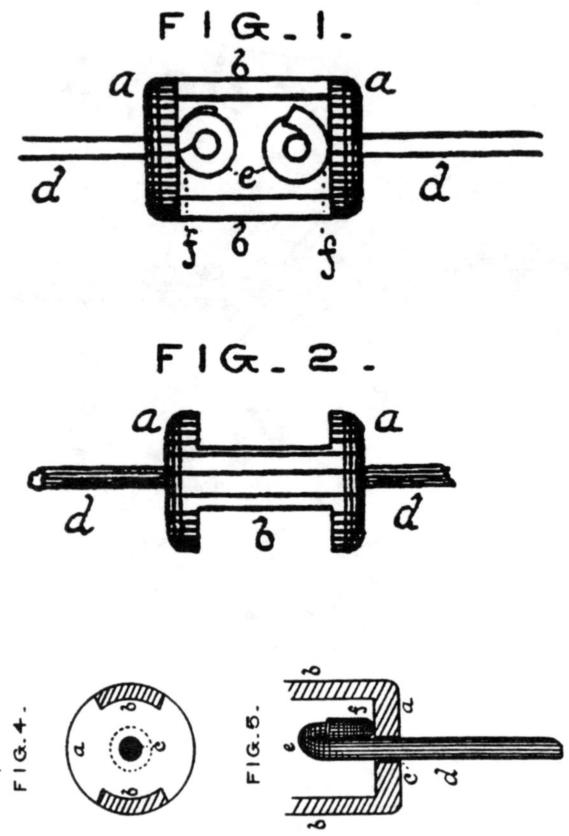

Clinton, DeWitt County, Illinois

The knob is a cast form per Figs. 4 and 5. The wires enter each end and are coiled as per Fig. 1. These end coils allow for freedom to pivot much as a universal joint.

September 25, 1883   Knot   285,764

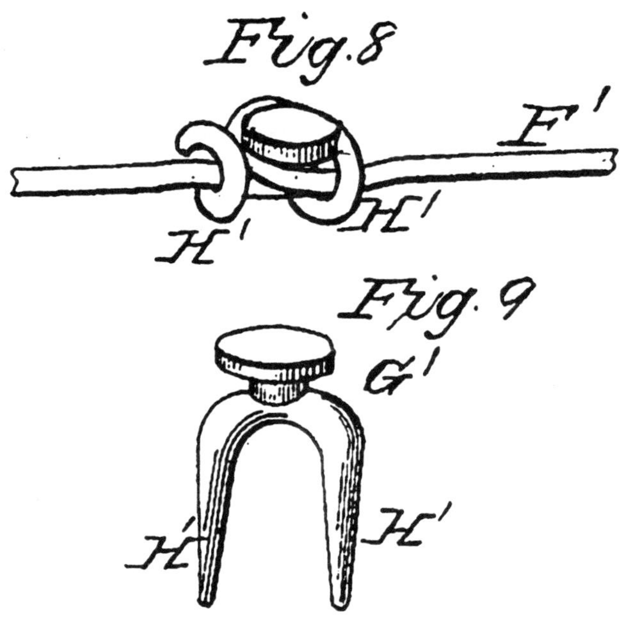

Rushville, Schuyler County, Illinois

"The check-wire as shown in Figs. 8 and 9 consists of a strong steel wire F' which, at suitable intervals, is twisted around the neck G' of a bifurcated piece of malleable iron or other suitable metal. The ends $H'H'$ are wound around the wire on both sides of the loop formed by the wire, forming a strong knot, which is not easily displaced, and can be made with little labor."

November 6, 1883   Throw-off Ball   288,158

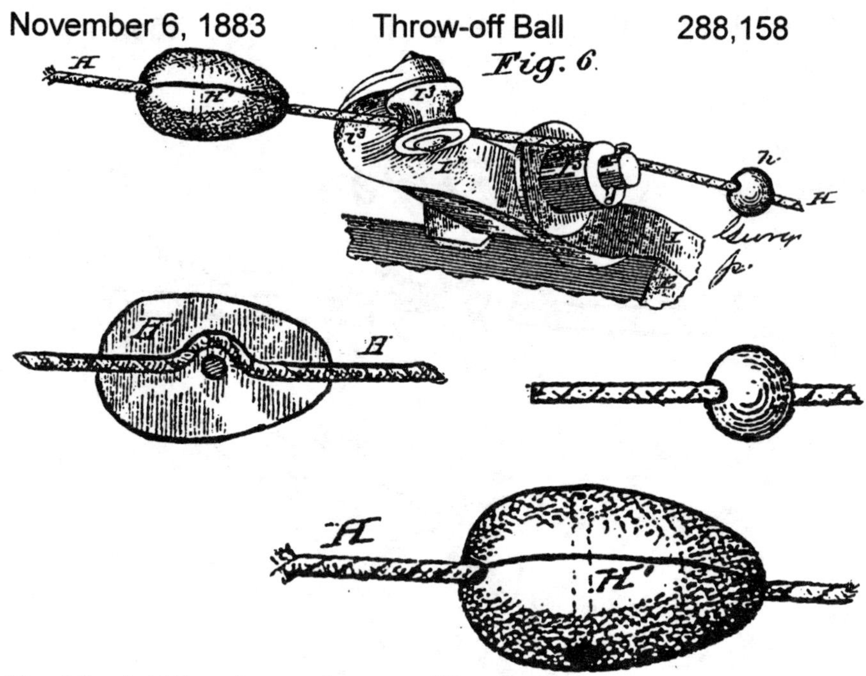

Rockford, Winnebago County, Illinois

"The object of the second feature of my invention is to provide for the automatic detachment of the button-wire or check-line from the machine when the latter has reached the end of its course. It consists in the combination of a ball or bulb on the wire, larger than the buttons, which operates to throw the wire off the pulleys. This feature is adaptable to any planter in which the button-wire does not cross the machine."

The large wooden bulb on a wire or rope with the smaller knot would be a rare find for the collector as there would be perhaps only one such bulb in a roll of check-line. The type of smaller wooden/metallic knot was not stated. The preferred shape of the throw-off ball is oval or egg-shape.

December 4, 1883   Knobs   289,468

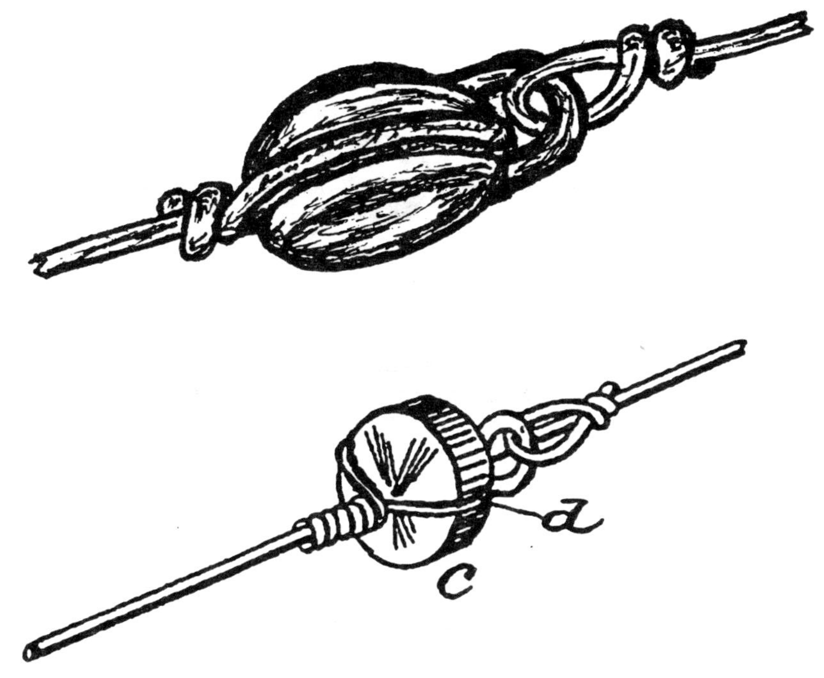

Monmouth, Warren County, Illinois

"The knotted check-wire consists preferably, of a number of knobs c, shaped like a shank button, having a groove d around the body, passing through the shank, in which the wire is placed, forming a loop, the end of which is twisted around the wire at the upper surface of the button, while the end of the other piece of wire is fastened to a shank of the button, thus forming a series of long links of wire, each link having a button."

The sketch of the shank-button is conjecture by the author. It's probably the 'padlock' cast as it was easier to make and still conform to the spirit of the patent.

February 19, 1884   Globular Knots   293,888

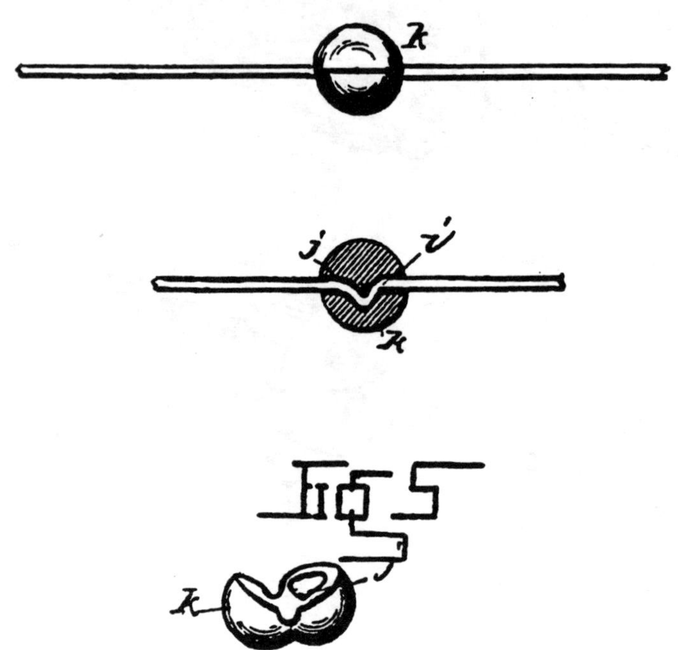

Ottawa, LaSalle County, Illinois

"At the point where the obstructions are located there are v-shaped bends or indentations i in the check-rower wire imbedded in grooves j of corresponding shape, in the middle of the obstructions k. These obstructions are of metal and are firmly clasped over the wire, and when finished are globular in form, as shown."

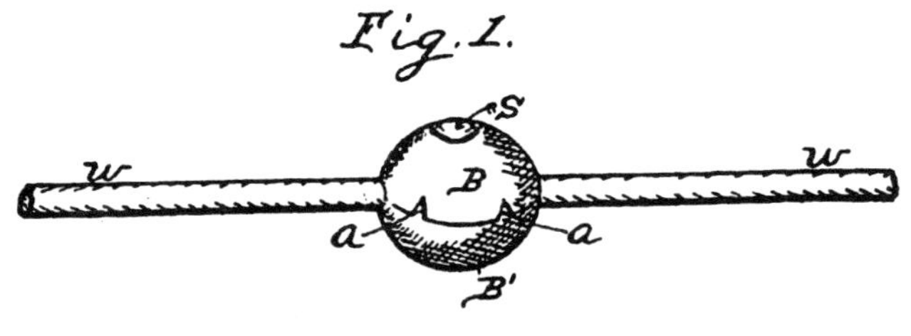

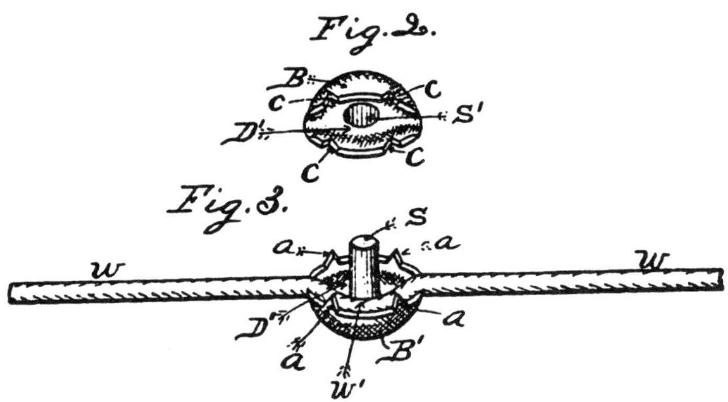

Joliet, Will County, Illinois

"The stop is composed of the two semi-spherical halves B & B'. The principle new feature in this invention is in serrating the meeting edges of said half-spheres B & B', or providing them with the projections or teeth a in one half and the corresponding notches c in the opposite half in such manner that when the two halves are united they will appear as shown in Fig. 1."

April 15, 1884   Stop   296,981

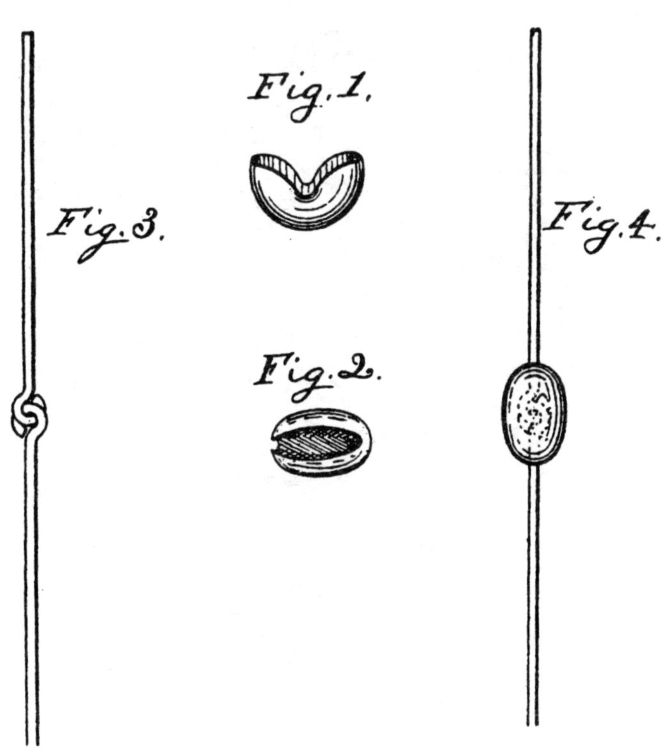

Joliet, Will County, Illinois

"To construct my improved wire, I cast or make a shell in shape like Fig. 1, of some suitable malleable metal. The shield or shell is then closed up, as shown in Fig. 2, leaving it sufficiently open at the edge to admit the looped wire as shown in Fig. 3, which is next inserted and the shell closed up, bringing its edges close together over the wire, as shown in Fig. 4."

April 22, 1884   Button   297,202

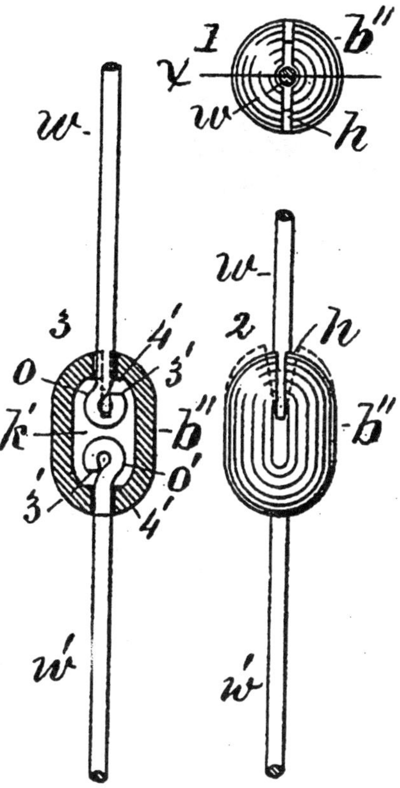

Springfield, Clark County, Ohio

"The wire of my check-rower is in sections, and connected by a hollow ellipse-shaped shell cleft through diametrically from one end to the center. The end of the wire is turned into a circular eye, which is turned on end and drawn inside the shell of the button b'' through the closed end in one section of the wire, and the eye in the other section is inserted through the cleft h and the latter afterward closed up."

April 22, 1884   Button   297,265

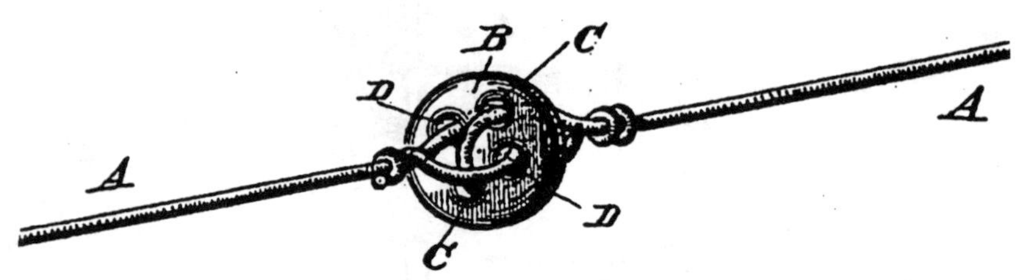

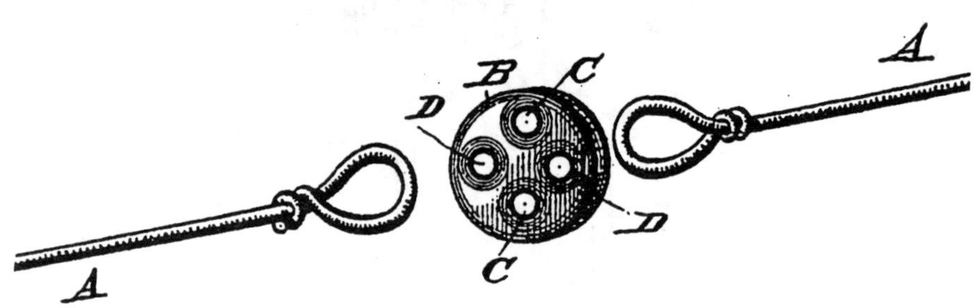

Sterling, Whiteside County, Illinois

"B is a washer or button designed to actuate the check-rower, and provided with the holes CC and DD, formed through the disk. The end of one of the sections A is passed through one of the holes C and returned through the other hole C and firmly twisted upon itself. The adjacent end of the next section A is passed in a contrary direction through one of the holes D, and back through the other hole D and in like manner twisted firmly upon itself."

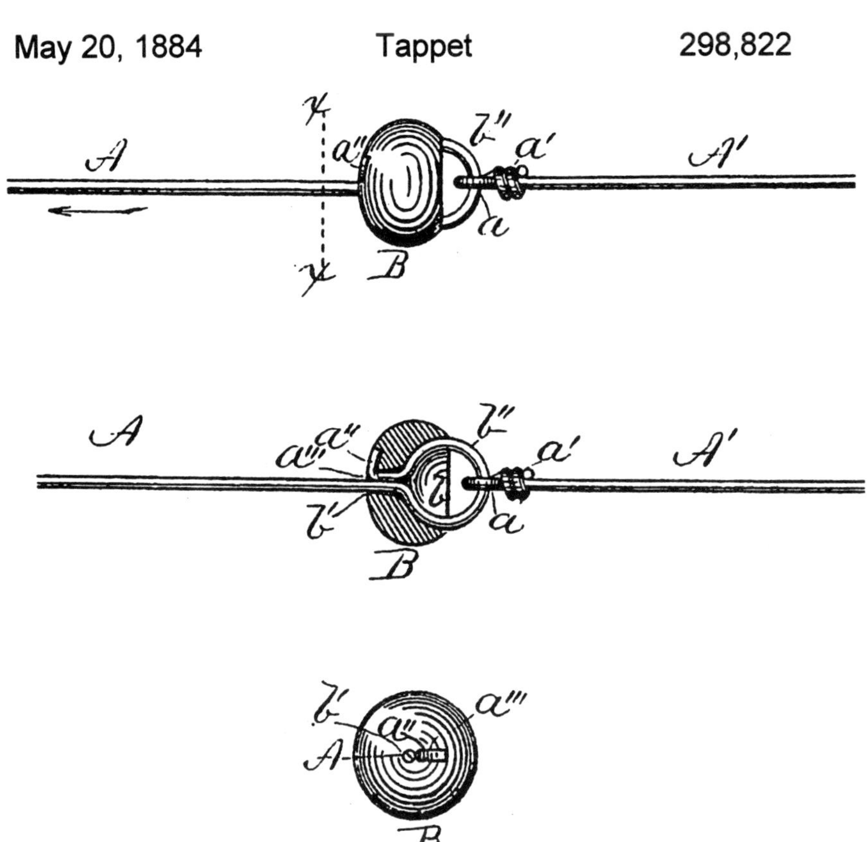

Galesburg, Knox County, Illinois

This patent is an improvement over 258,217. He said that patent is objectionable because the wire is coiled around itself and can trip the fork lever too soon.

"The object of my present invention is to remedy this defect; and consists of a loop on the end of the section of wire passed through the tappet and secured therein by bending the end of the wire down into the recess or cavity in the tappet."

July 8, 1884   Knots or Knobs   301,825

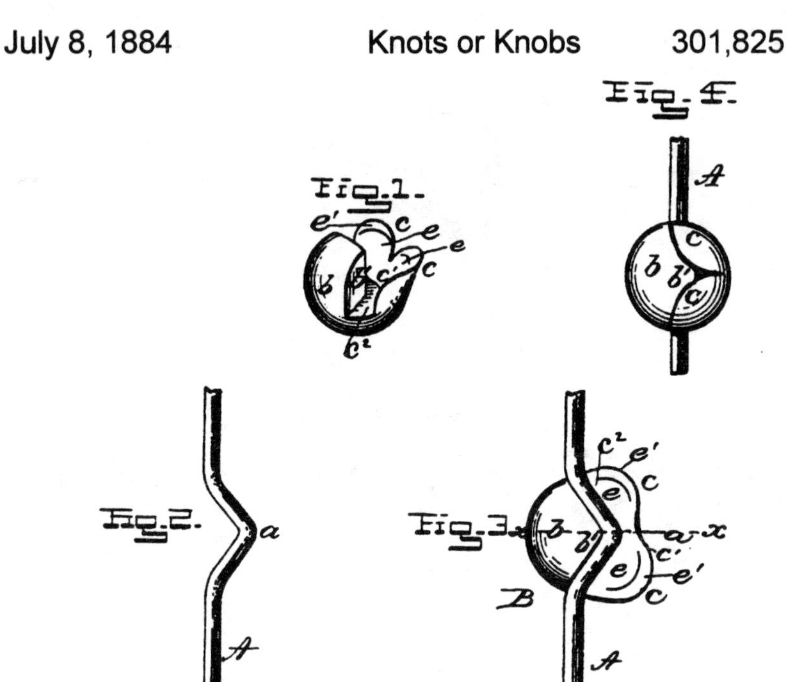

Joliet, Will County, Illinois

"B designates one of my improved knots, or knobs, which I prefer to make of cast malleable iron. Each knob B is cast with an angular shoulder b' formed on the body b and also with two lips cc separated by an angular notch c' as clearly shown in Figs. 1, 3, and 4. Between the angular shoulder b' and the lips cc is left a deep groove c'' which is also angular and adapted to receive the bent or crimped portion of the wire A. After the wire is positioned in the groove c'' the lips are hammered down and closed over the wire thereby securing the knob on the wire."

July 22, 1884   Plaster   302,514

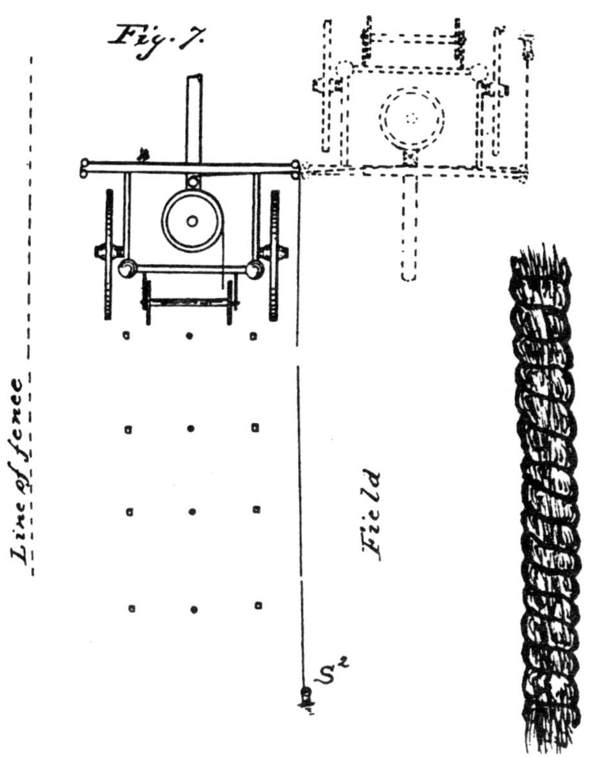

Cincinnati, Hamilton County, Ohio

"The object of dropping the plaster or marking material in line with the two hills of corn is to have it serve as a visible guide for the operator on the return trip of the machine and to enable him to align accurately succeeding rows of hills with those previously planted."

The check-rope is smooth. Various check spacing is accomplished by the variable pulley, Fig. 3.

August 5, 1884   Tappet   303,169

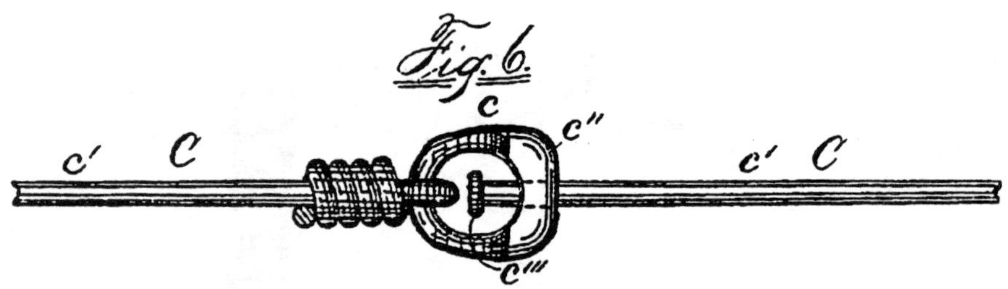

Galesburg, Knox County, Illinois

"My tappet wire, with tappets or knots c, is formed as shown in Fig. 6. The sections c' of wire are united by the tappet c which is a link with an enlarged end c'' through which the end of one section of wire passes and has a head c''', formed thereon, which holds the wire therein and forms a swivel. The end of the other section of wire passes through the other end of the link or tappet c and bent back upon itself and secured by coiling the end upon the main wire so as to form a universal-joint connection between the sections of wire and, also a swivel to permit flexure in any direction."

September 9, 1884   Button   304,812

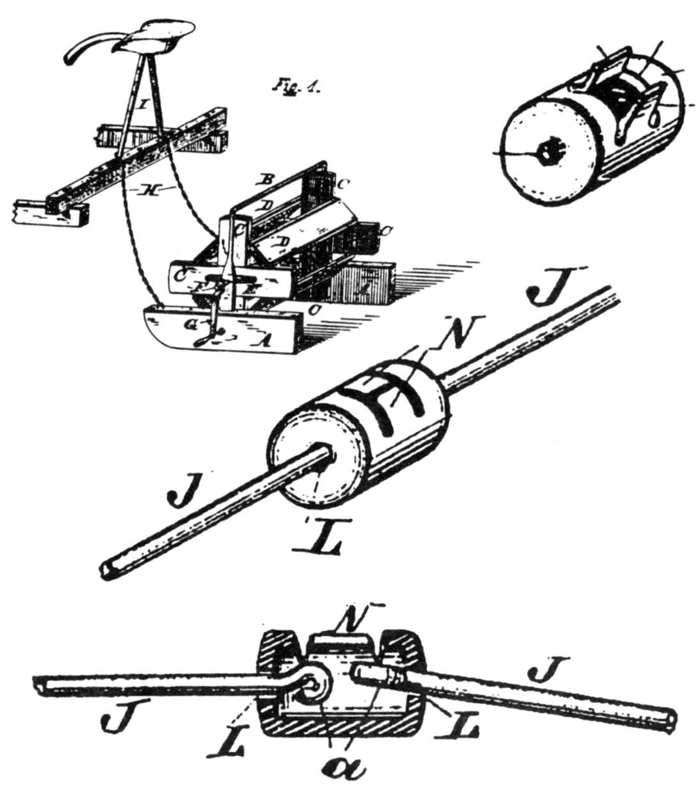

Springfield, Clark County, Ohio

Claim 5: "An operating button for check-row wires, consisting of a hollow shell with solid centrally-perforated ends and an opening upon the side adapted to be closed by one or more integral lips, in combination with the links of the wire adapted to be swiveled therein, ..."

This patent includes the reel per Fig. 1.

November 11, 1884   Button   307,755

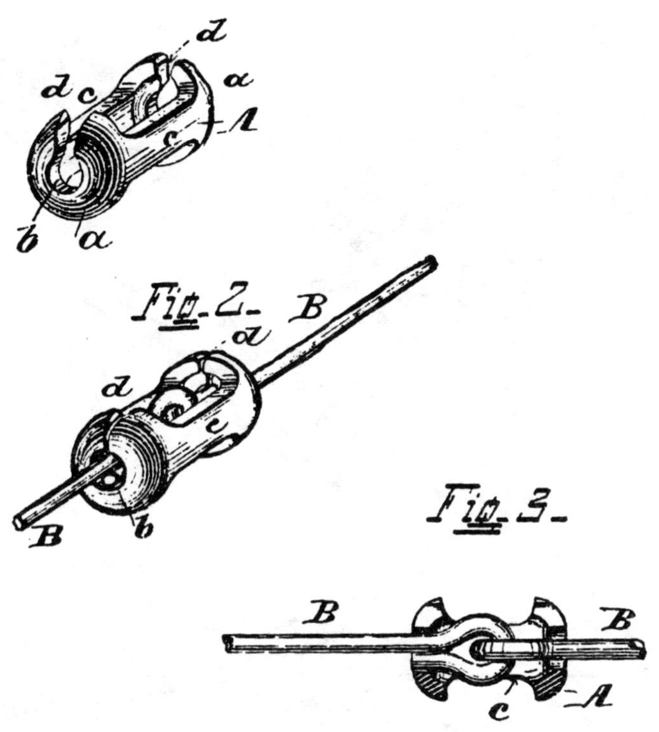

Springfield, Clark County, Ohio

"The button A is preferably a malleable shell with rounded heads a containing perforations b for the passage of links of wire. It is cut away on each side leaving the connecting pieces c and has slots d through each head in line with each other and opening into the perforations b. These slots are open enough to receive the coupled ends of the links, after which they are closed, per Fig. 2."

NOBLE E. CROTHERS

February 24, 1885   Knot or Ball   312,977

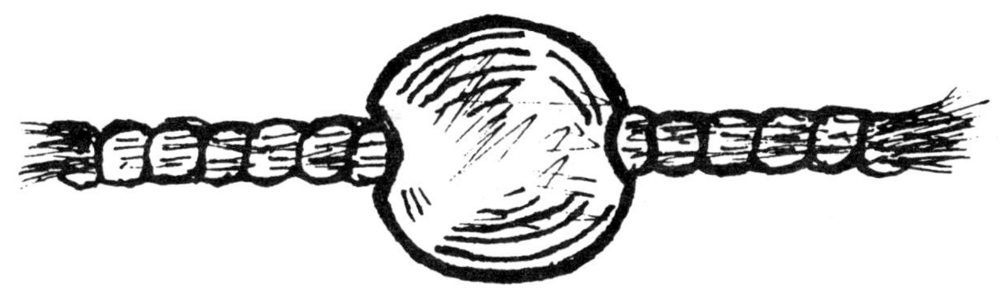

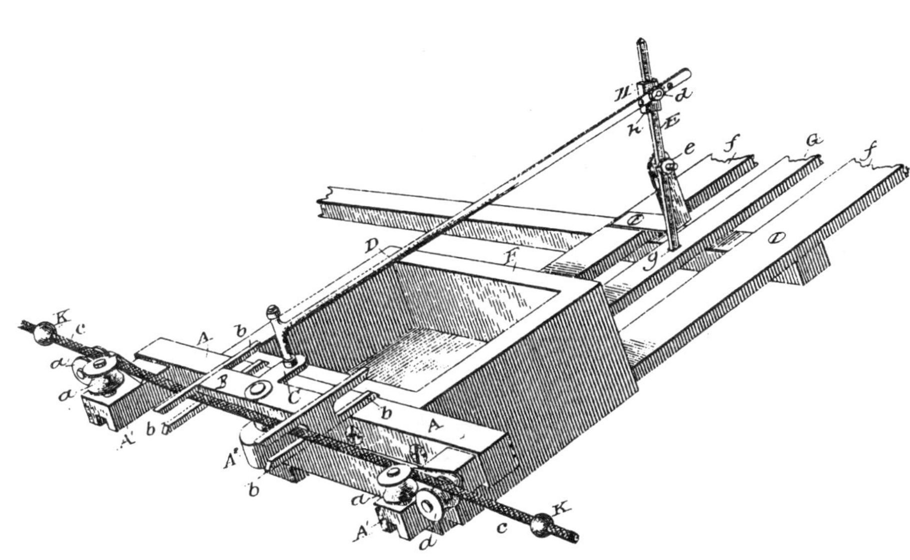

Paxton, Ford County, Illinois

Though what appears to be a round wooden knot, no claim
was made for the check-line. Probably Bosworth 254,267.

April 28, 1885 — Stop — 316,792

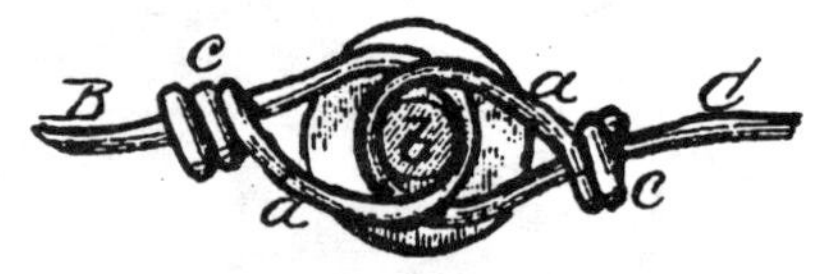

Sterling, Whiteside County, Illinois

"The form of stop *a* may be variously modified so long as it will admit of turning on its axes in a direction lengthwise of the line to which it is connected."

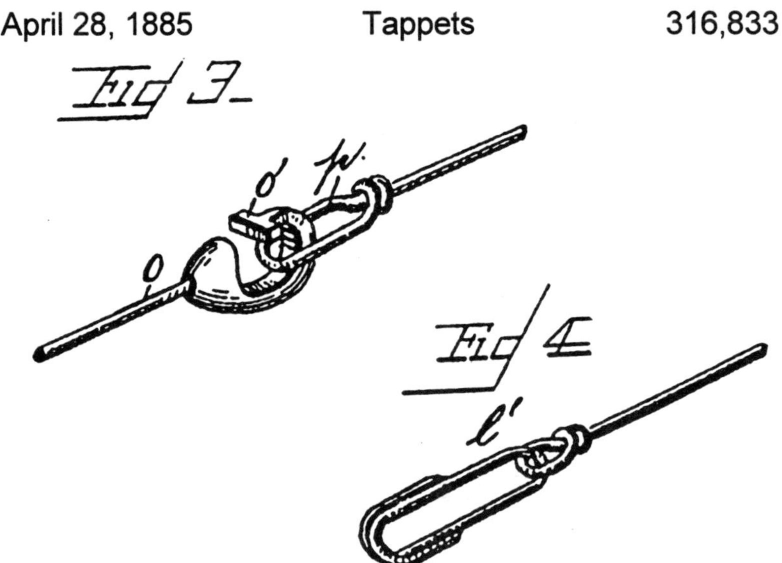

Riverside, Washington County, Iowa

"In practice, the check-row cord is made of short sections of wire between which are secured the tappets as shown in Fig. 3, said tappets having one end conical, which is provided with an opening through which passes the end of the wire o, which is looped upon itself, this end of the tappet being recessed for the reception of said loop. The body portion of the tappet is curved upwardly, as shown, its terminal portion being provided with a cross-bar o' which is located below the plane of the opposite end of the tappet. The opposite end of the wire is provided with a loop p which is of sufficient length when placed at right angles with the tappet to cross over the cross-bar o' so that it may be removed therefrom. By providing a tappet of this construction, sections may be added or removed from the line."

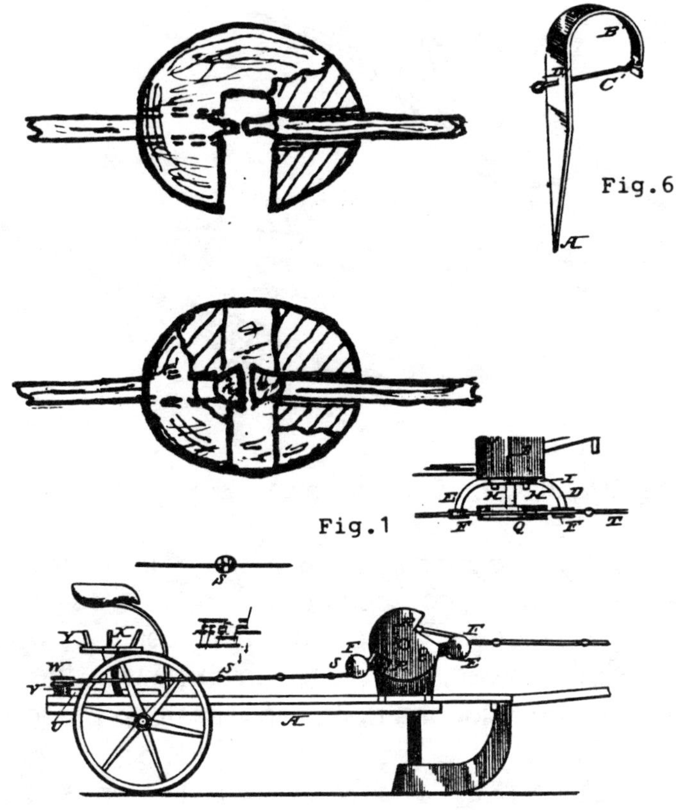

Fig.6

Fig.1

Bronson, Bourbon County, Kansas

This patent does not explain the tappet design.

It appears there is a slot around the tappet in Fig. 2 and across it in Fig. 1. The tappet must be metal.

Fig. 6 shows the anchor used with the wire.

HENRY PEARSON AND GEORGE JARMIN

May 26, 1885   Double Knobs   318,650

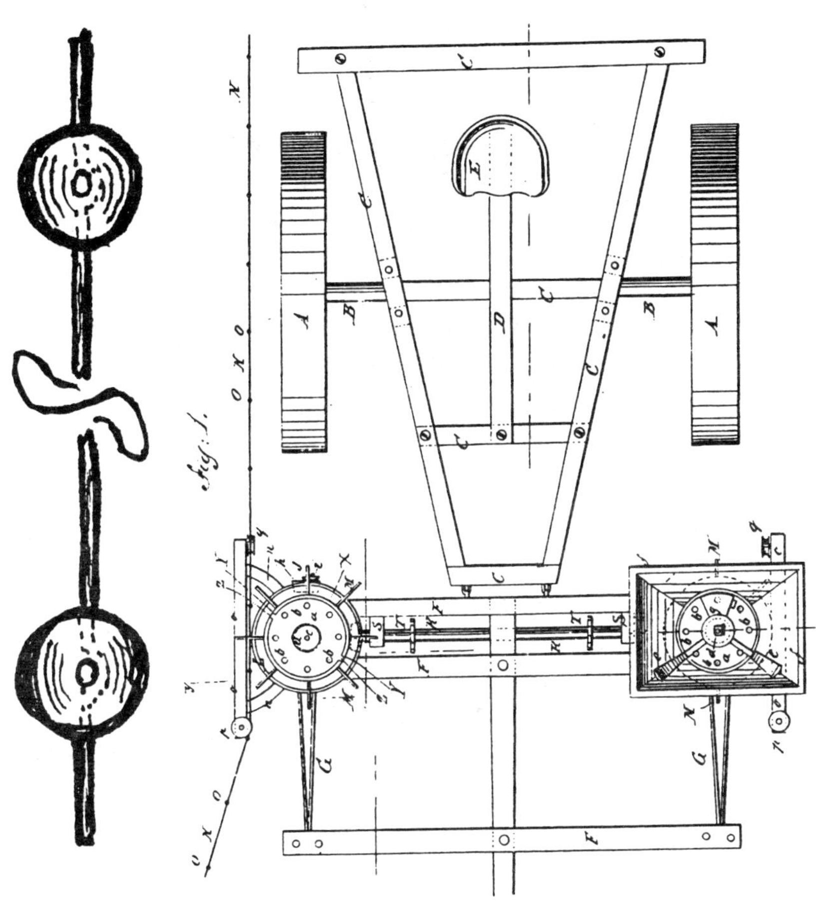

Genoa, Nance County, Nebraska

This patent calls for use of knobs or balls close together to reset the seed drop. This is hill half-distance between balls.

May 26, 1885   Knot   318,874

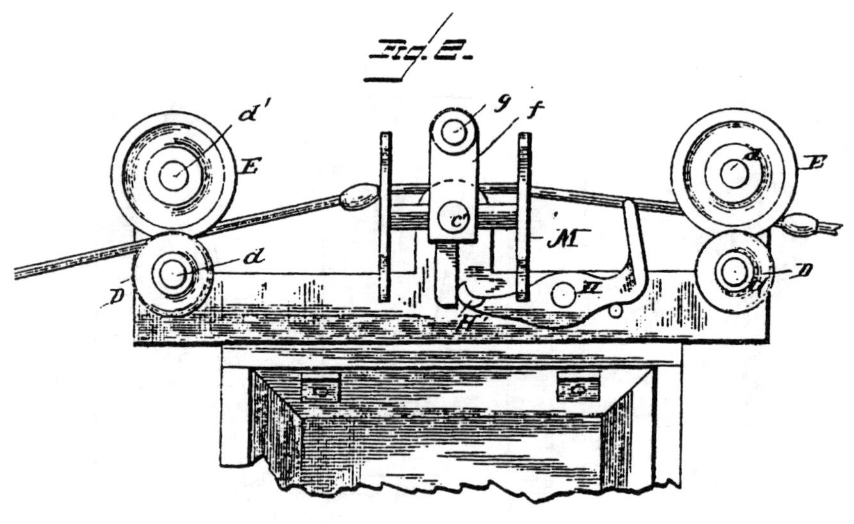

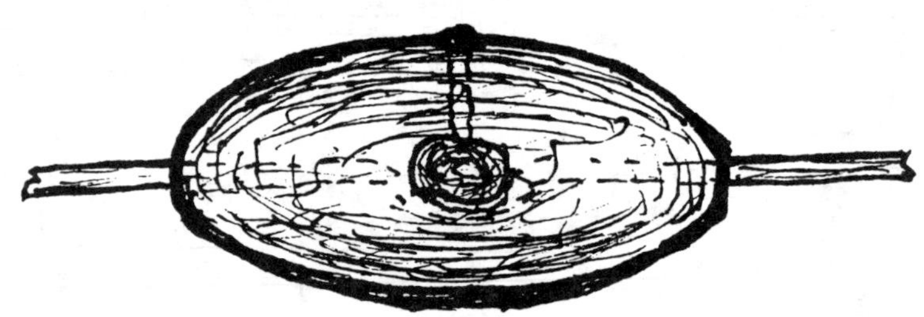

Paxton, Ford County, Illinois

The knot here is not claimed. However, it is shown here as it appears to be a longer knot than his round knot in 312,977. It is curious that he did not show that knot here.

October 20, 1885   Knots   328,452

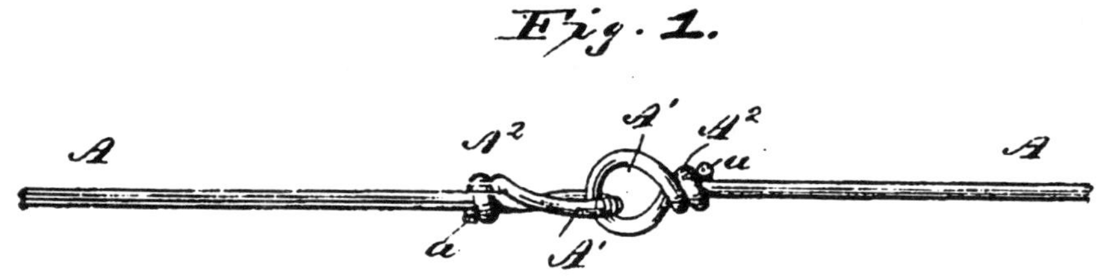

Quincy, Adams County, Illinois

This patent is an improvement over earlier knots such as shown in Fig. 1, where the wire is wrapped such that the end is apt to get caught and bent or broken, or trip early.

"To overcome this defect my invention as in Fig. 2, coils the end back over itself and presses it down behind the reversing coil, not only shielding the end, but providing a larger knot."

January 5, 1886   Knot   333,699

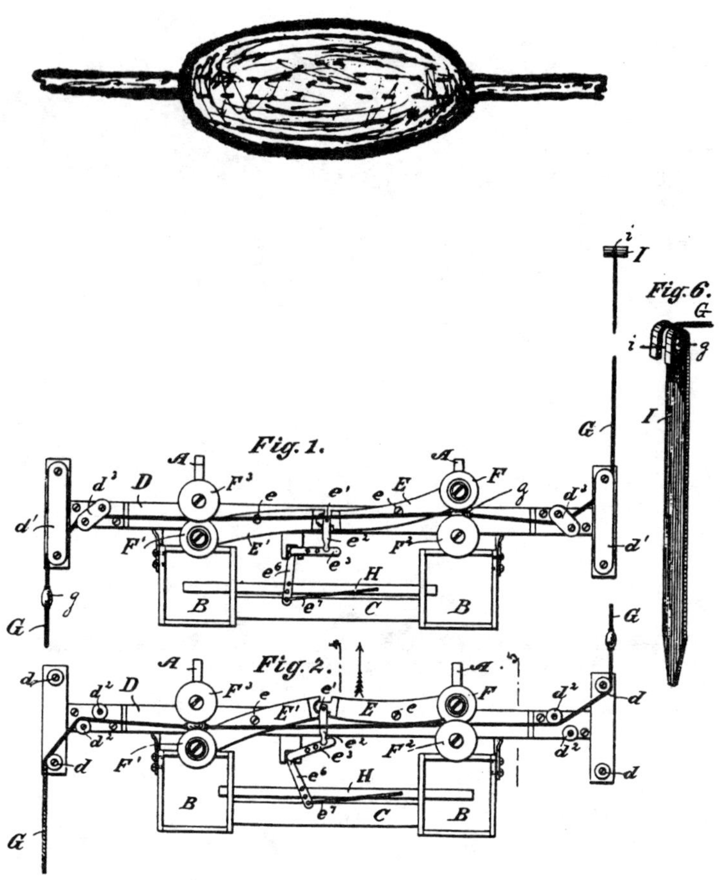

Aledo, Mercer County, Illinois

This patent is for an improvement on patent 223,190 that included his cast knot on a wire. The improvement is on the lever arrangement. However, note the knot g in Fig. 1. The knot is oblong. Fig. 6 is for an anchor included in 333,669.

April 13, 1886   Throw-off Disk   340,066

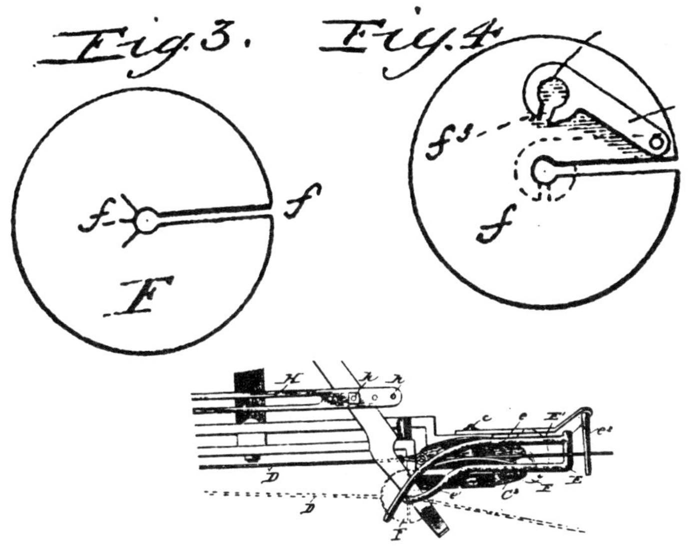

Pierceton, Indiana

"Fig. 3 consists of a thin metal disk having from its periphery to its center the slot or opening f narrower than the diameter of the check-row line, and having the center opening f' of a size larger than the wire. The wire is inserted by laterally separating the edges of the slot to permit the wire to pass to the center. Fig. 4 shows the disk with a supplemental latch F'. The slot on this disk may be larger than the wire, so the wire may pass freely to the center. The latch then retains it."

The purpose of the disk is to disengage the check-row wire at the end of the row without the need for the operator to dismount the planter until turned. The disk works in conjunction with attachment per Fig. 1.

October 12, 1886   Button   350,591

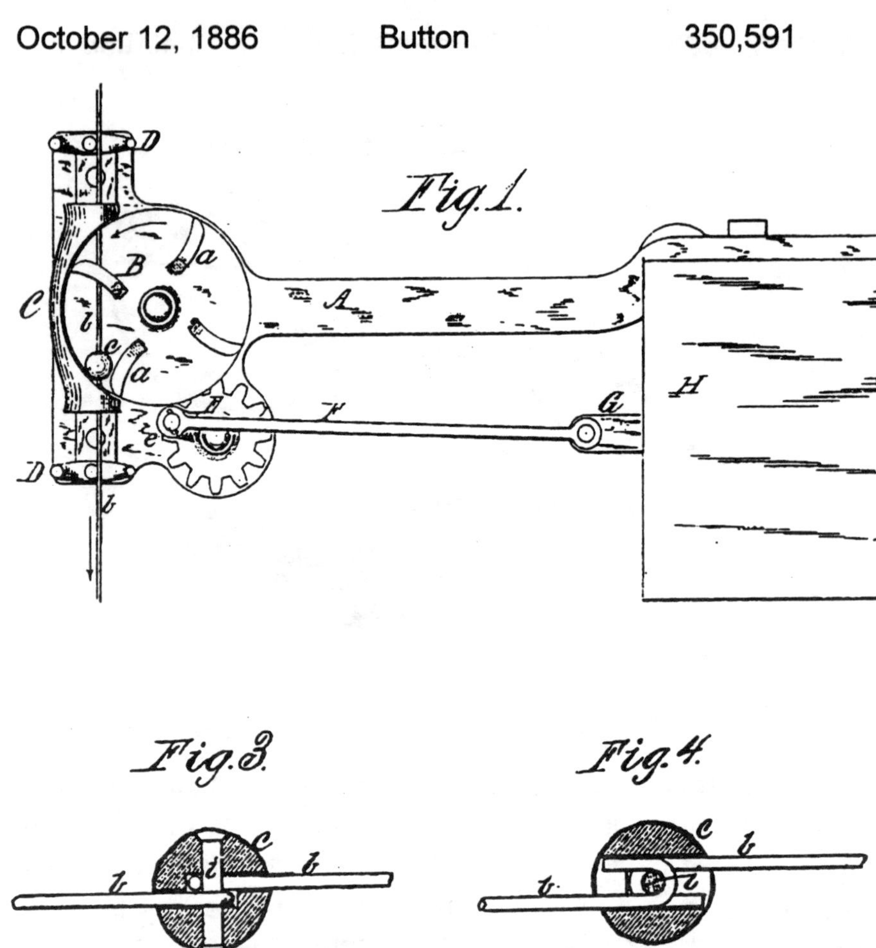

Monroe, Linn County, Iowa

"Through the center of the button in one direction extends a slot wide enough to receive the hooked ends of the wires bb. In the opposite direction, and through these hooks, passes the pin i, locking them together. To prevent the hooks from slipping out of the button back of the pin an offset is formed in the slot, as shown."

April 5, 1887   Swivel-Joints   360,825

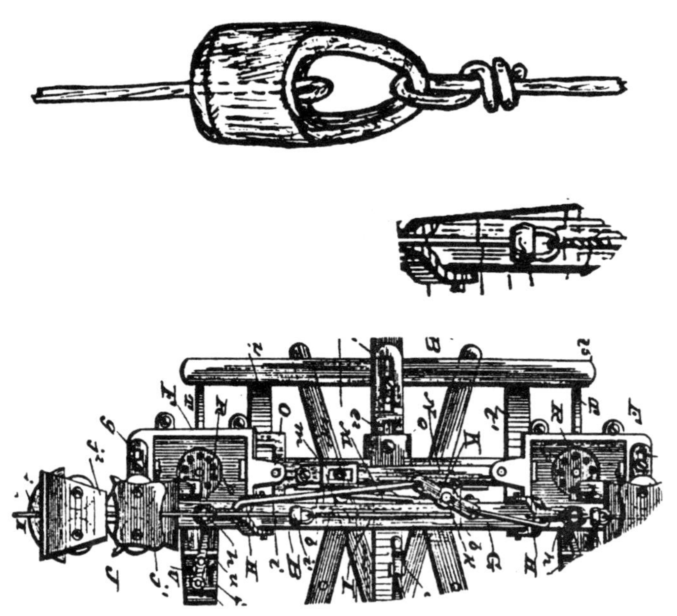

Rockford, Winnebago County, Illinois

" *I* is a wire composed of equal links *i*, of suitable length, united by swivel-joints. These joints are each composed of the hemispherical button *i '*, provided with a bail to which the end of a link is attached by a twisted loop, while the end of the adjacent link passes through an opening in the base of the button and is bent up, so as to be retained within the interior hollow of the same." This design is covered in Claim 3. However, no separate figure was provided in the patent. The above enlargement is from the planter assembly view.

Patent was assigned to Skandia Plow Company.

November 8, 1887   Tappet   372,671

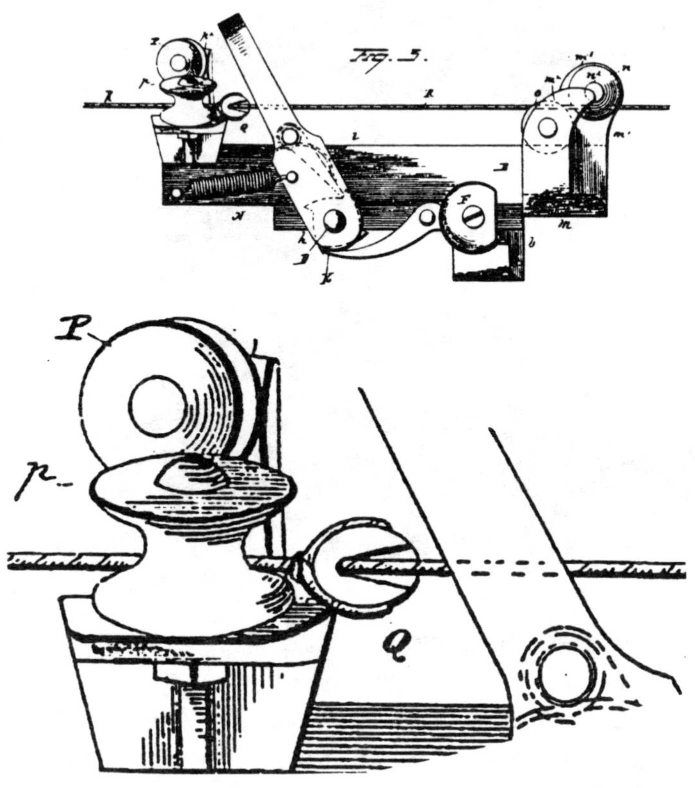

Tallyrand, Keokuk County, Iowa

The tappet shown as Q is not described. It appears to be shaped like an eggshell with another piece fitted inside to pivot. The wire or cord is knotted inside the internal shell and outside the external shell. This knot is operational in one direction only.

December 20, 1887   Balls   375,275

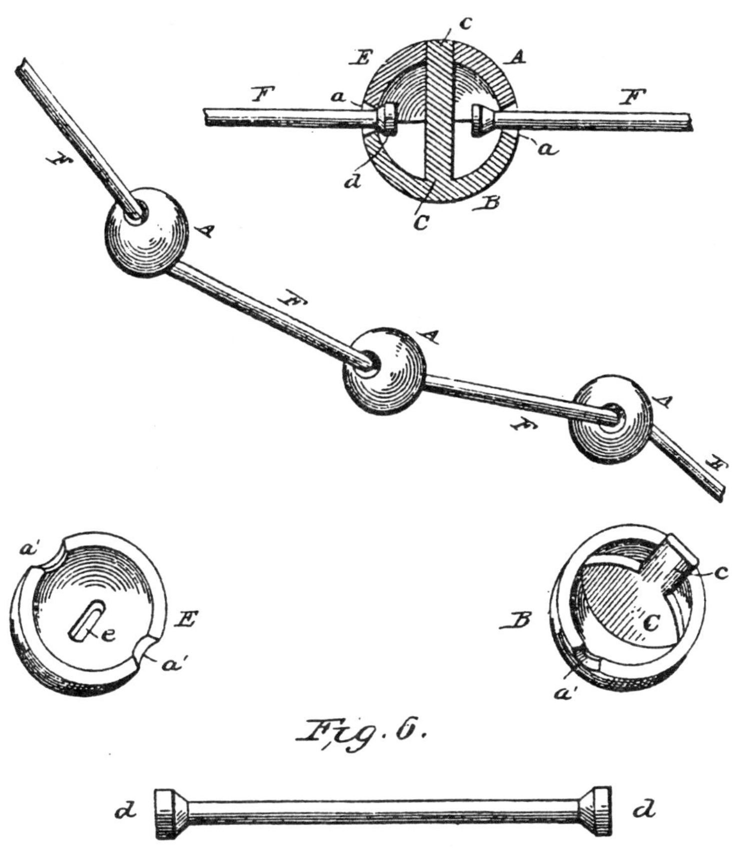

Peoria, Peoria County, Illinois

"My invention has for its objects to provide a flexible swivel-jointed link-wire the knobs of which are of perfect spherical form. The knob or 'balls' as I call them, are so constructed as to prevent the ends of the wires from touching. The links will not have any twists or bends."

147

AUSTIN S. AND EDWIN HOUCK

April 30, 1888   Knob   381,073

Bedford, Taylor County, Iowa

This patent shows a classic check-row planter. The gentleman sitting no doubt expresses his desire once the crop is harvested. The check-row wire appears to have a tapered wooden knob.

February 11, 1890   Knot   420,992

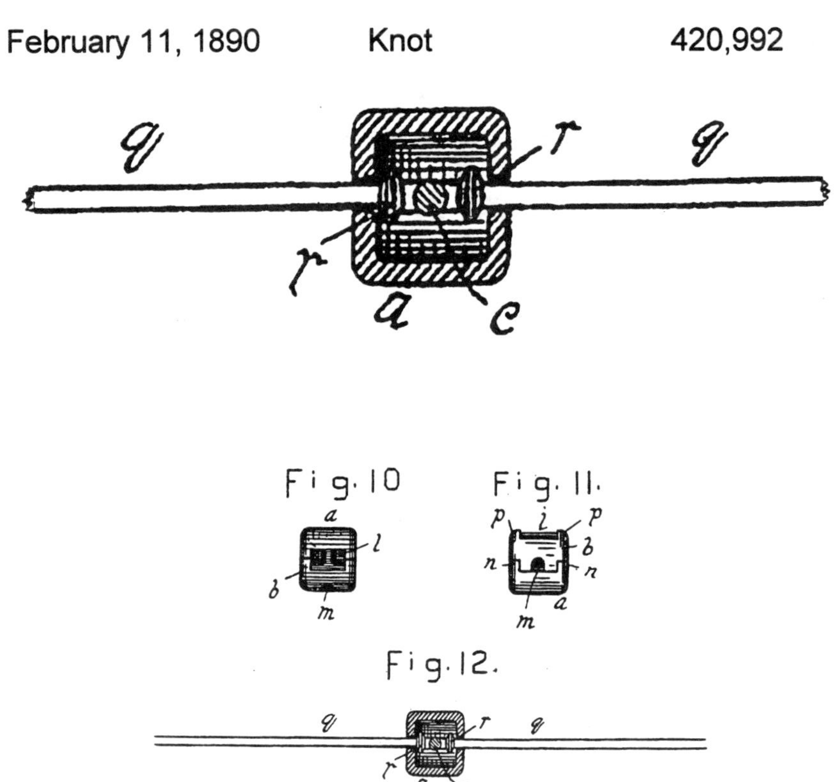

Decatur, Macon County, Illinois

"The object of my invention is to provide stops that will hold sections of wire together and that may be readily connected and disconnected to pass trees and other obstructions, to adapt the wire to the length of the field, and to replace worn sections with new ones."

"To make or break a connection in the field, a stop is held with an implement, the jaws of which are preferably inserted in recesses l and m, and the head is turned into the desired position by a wrench or pinchers."

September 30, 1890   Tags   437,239

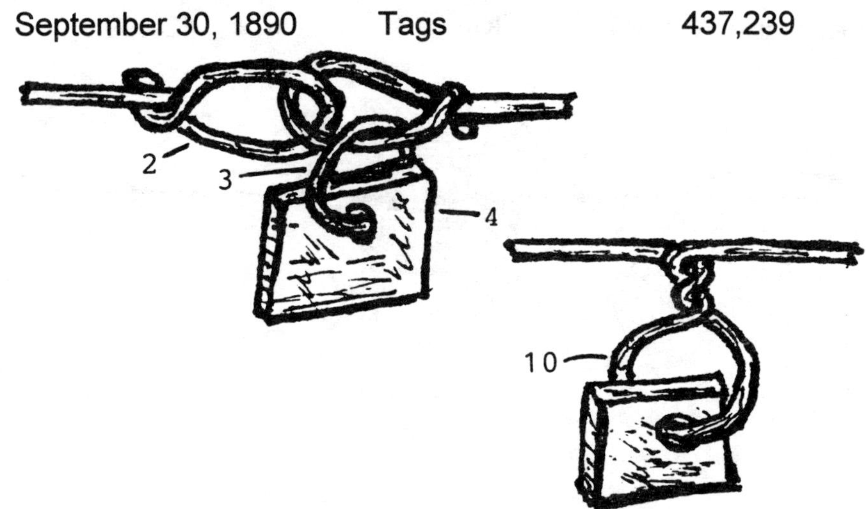

Brownsburg, Hendricks County, Indiana

"This invention relates to check-wires for corn-planting; and it consists in what we designate as a 'hand-planting' 'check-wire' for the purpose of indicating the exact spot where the seed shall be deposited when ordinary hand-planters are used, thus insuring regularity in the planting of the seed."

"Our improved check-wire may be constructed of a series of links, 11, of equal length, <u>made of ordinary binder-wire or other wire</u> of suitable dimensions and quality, each link or length being provided at its ends with loops or eyes 22, by means of which the said links are joined flexibly together. The links are each to be of a length corresponding with the distance between the hills of corn. At each junction between the two links *I* attach by means of a ring or loop 3 <u>a tag 4, which may be of sheet-metal, leather, or any other suitable material</u>, and which is preferably of some bright or conspicuous color, which will enable it to be seen very readily. . . ."

September 30, 1890　　　Tags　　　437,239

Check-Wire for Hand-Planting

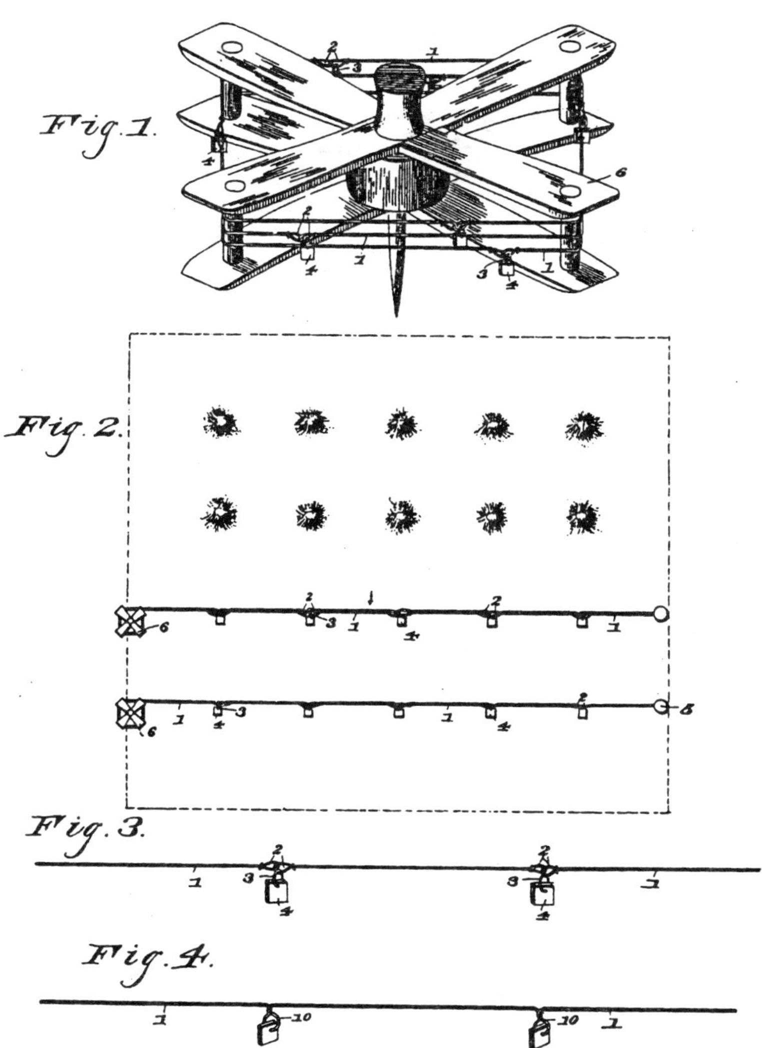

Brownsburg, Hendricks County, Indiana

December 2, 1890   Button   442,032

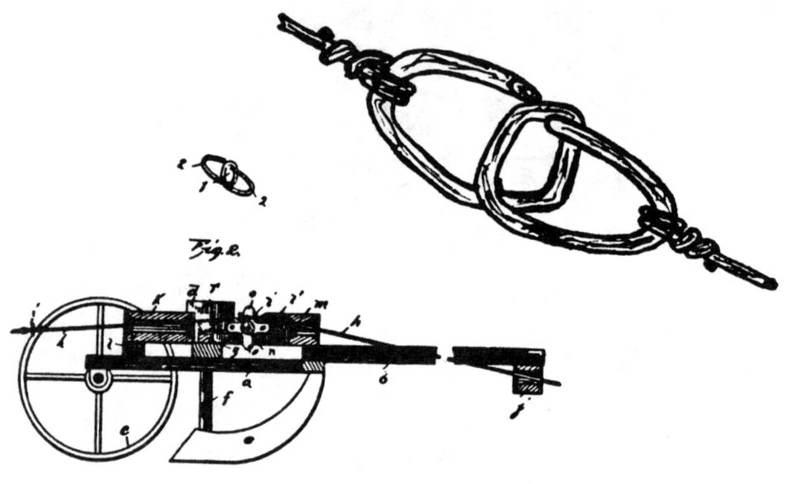

Above sketch is found variation

Vail, Crawford County, Iowa

"Each button consists of a single piece of wire, bent near its center to form the open circle *1*, and the ends of the wire are bent in opposite directions to form the loops *22*, extending in the opposite directions from the open ends of the central circle. The ends of the wires *3* are secured loosely to said loops *2* by means of eyes, so that the buttons act as links to unite the sections of and form the complete wire."

May 5, 1891   Button   451,871

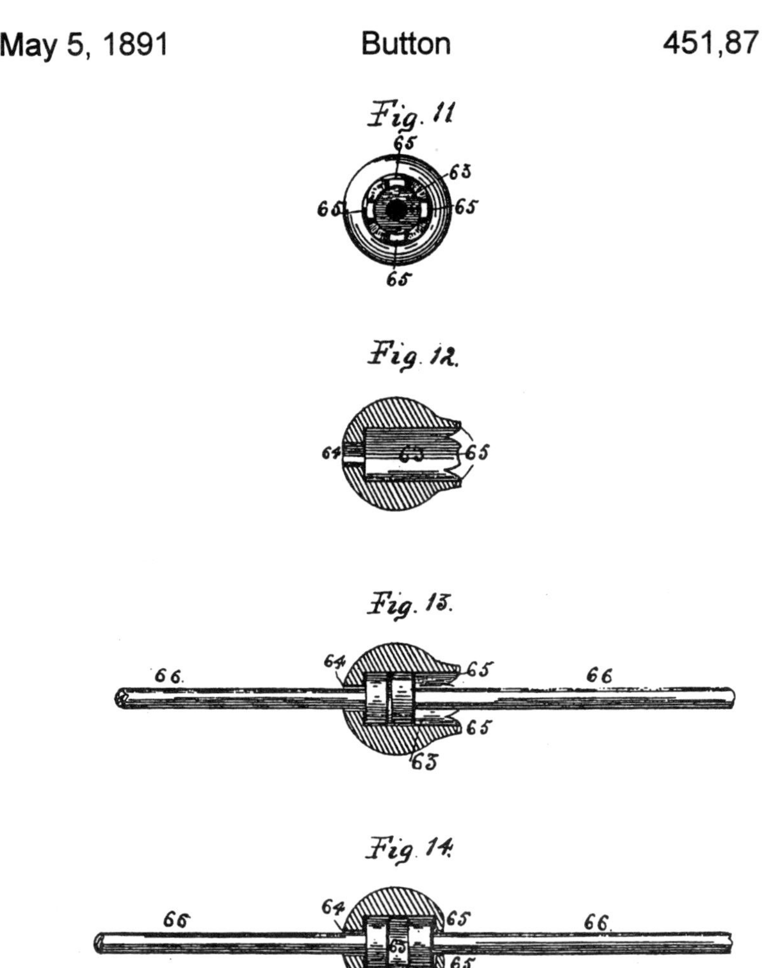

Rockford, Winnebago County, Illinois

"Figs. show an improved button which is spherical in form, the headed wires are inserted and the lips compressed to form the completed button."

Fig. 1 shows a reel including a hand-crank.

February 23, 1892   Stops or Buttons   469,403

Chicago, Cook County, Illinois

This is an improvement over patents 208,814 and 236,025.

"In carrying out my invention, I form the stop or button A in one piece, and while said button could be struck up, I prefer, as a matter of decided improvement, to form it by casting."

October 30, 1894   Buttons   528,235

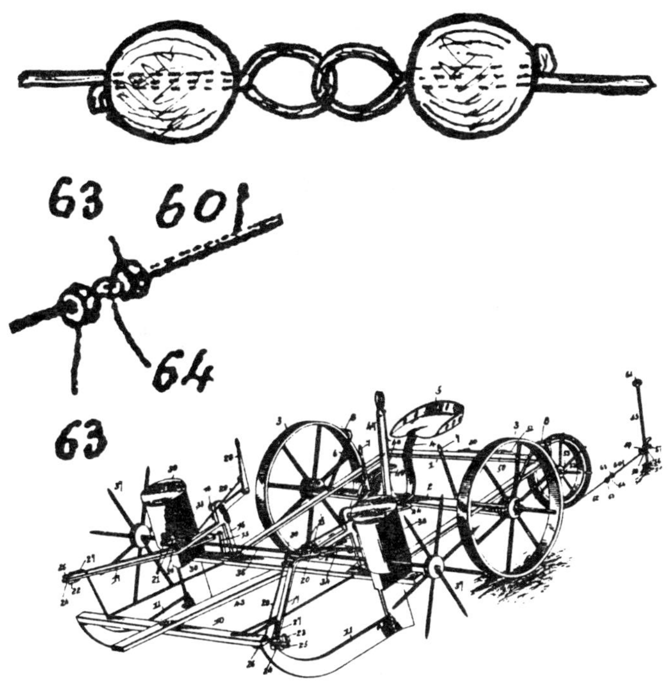

Viola, Mercer County, Illinois

"The check-row wire may be of any suitable formation or construction, and I have herein: illustrated a construction, which I prefer to employ. The wire consists of a series of short sections *62*, the ends of each section passing through spherical buttons *63*, and beyond these buttons, the ends adjacent to the buttons are given a twist to form eyes *64*, the eyes of one section interlocking with those of the other, whereby a continuous wire is produced and one in which the buttons will not catch with the bifurcated check-row levers."

April 9, 1895   Ball Knot   537,261

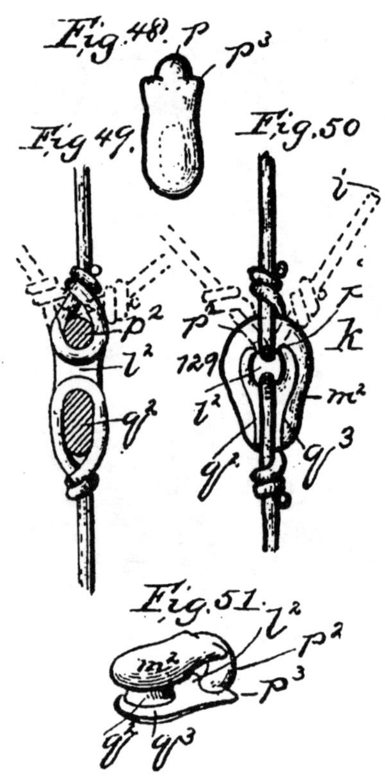

Peoria, Peoria County, Illinois

"A ball or joint of this sort is of superior nature for while it allows all the necessary flexibility, it prevents the wire sections from kinking either on the reel or on the ground, and the wire passes smoothly to the check-row levers no matter how it may be twisted, the ball or joint being of such a character as to insure that the wire sections shall be in line when they reach the levers."

March 16, 1897   Check-Row Wire   578,830

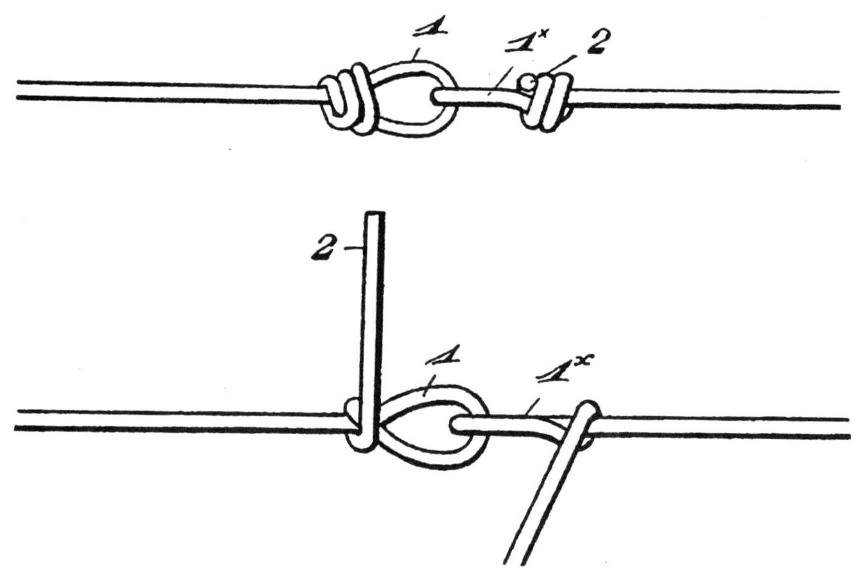

West Pullman, Cook County, Illinois

"My invention relates to improvements in check-row chains or wires; and the object is to provide a novel eye or loop connection or fastening for the meeting ends of the links of the check-row wire which will unite the ends of the links in such a manner that the check-row wire may be drawn from the reel or spool with certainty and expeditiously, and the checks will engage the forks .or guides in such a manner without damage to the forks or guide on the reel."

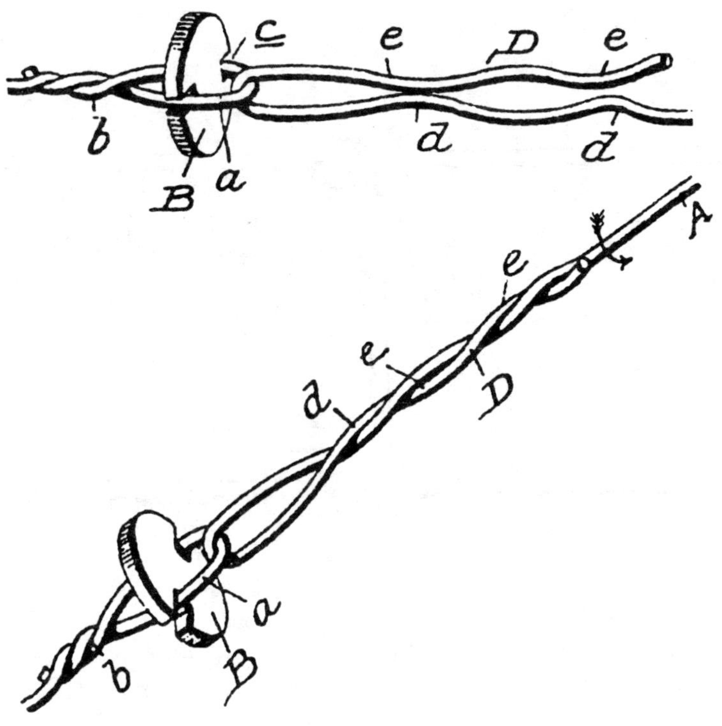

Streator, LaSalle County, Illinois

"My invention relates to improvements in check-row lines; and it has for its general object to provide a cheap and simple check-row line made up of a plurality of sections, embodying such a construction that sections may be added or removed to lengthen or shorten the line without the aid of any implement."

The main portion of the line per Fig. 1 has two or more kinks to facilitate the easy connection or removal of sections.

July 18, 1890   Hook and Tappet   628,959

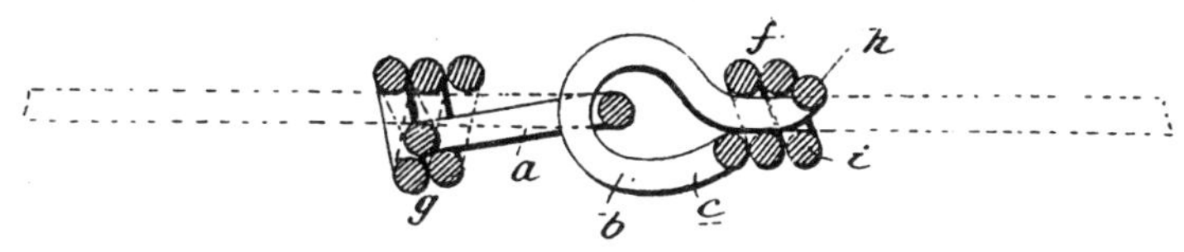

Chicago, Cook County, Illinois

"My invention relates to improvements in wires for check-row planters; and it consists mainly in the formation of the <u>hook and tappet integral</u> with part of the wire. The hook being so formed that it may be quickly and conveniently opened and closed to place or remove a section of wire."

This patent refers to being an improvement to 578,830 and it can be used on wires other than check-row.

July 21, 1903   Stops /Knots   734,101

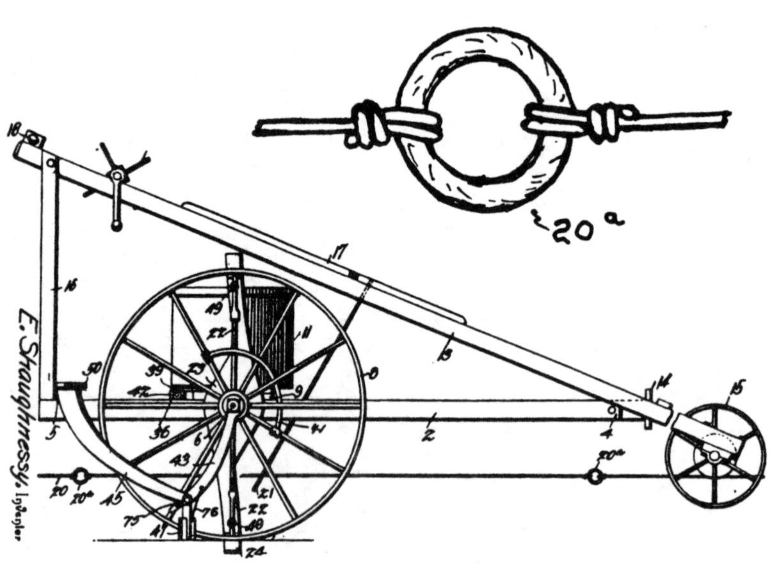

Tomah, Monroe County, Wisconsin

"These stops will preferably be rings inserted into the wire, the object being explained: The function of this rod, *21*, is to denote to the operator whether the seeds are being dropped regularly or not by 'clicking' upon the spaced rings on the wire, and if the clickings do not correspond to the actions of the planting means, then the machine can be stopped and the rear wheels raised and moved forward or backward until the relative position of the stops and discharge tube correspond and the planting continued. Thus the regularity of the planting and the uniformity of the rows or hills will be maintained."

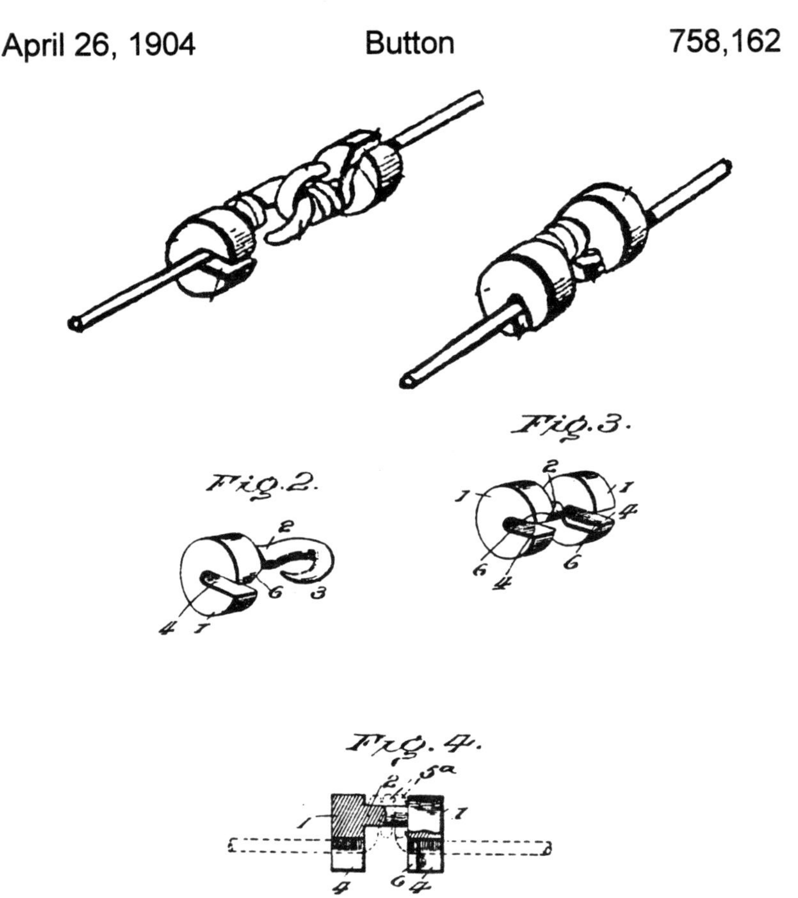

San Jose, Mason County, Illinois

"This invention provides a button for planter check-lines having a passage for the line and a reduced portion for the line to wrap about to fix the position of the button, and yet have the wraps come wholly within the confines of the button, so as to not come in contact with the fork or other operating part of the planter mechanism."

May 24, 1904 Check-Ball 760,514

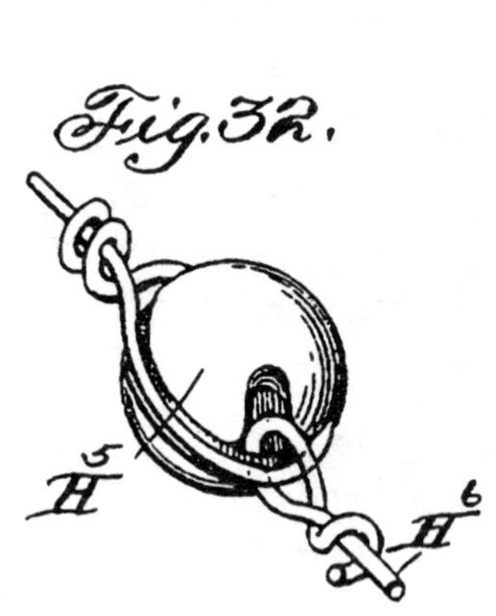
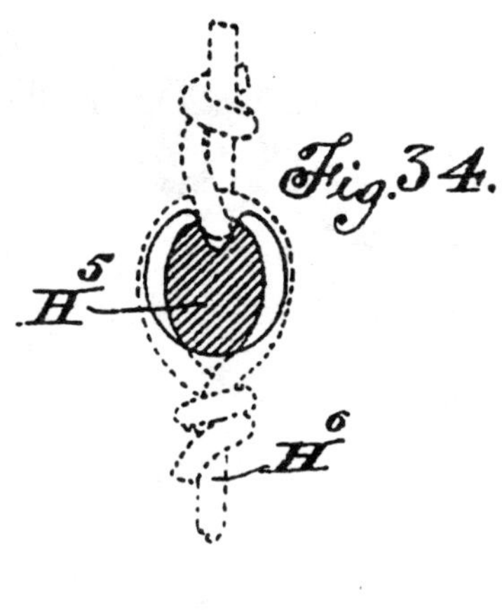

Peoria, Peoria County, Illinois

This patent includes the check-ball and a reel with the planter. The ball per above Figs. 32 and 34 look very similar to the ball in Berrien's 246,273. However, close inspection of the section view in Fig. 34 indicates the recess for the wire $H6$ does not circumvent the ball completely. It fades out tangent to the ball. The ball is round, but the groove is oblong.

Section view of Fig. 11 of Berrien shows a concentric groove. The difference is minor but obvious.

The patent was assigned to Avery Manufacturing Company of Peoria.

September 24, 1907   Tappets   866,708

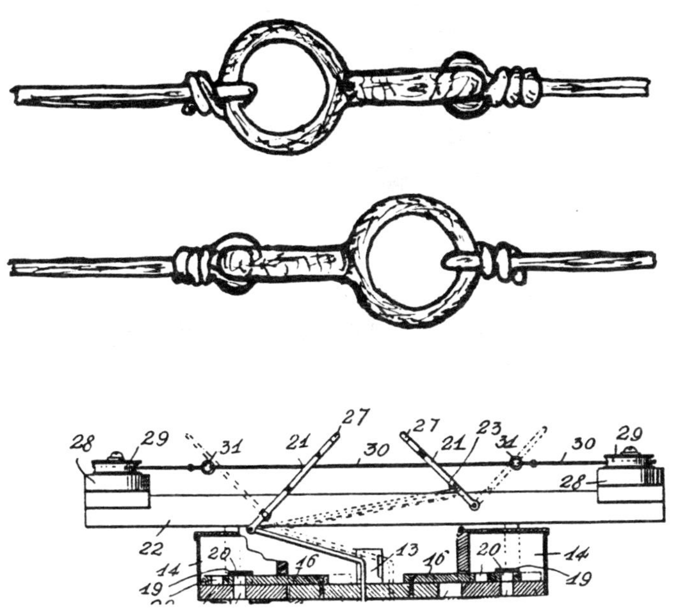

Wichita, Sedgwick County, Kansas

The tappets appear to be double rings formed in one piece at 90 ^{0} so one ring passes and the other actuates the lever. The tappet is alternated on the rope per Fig. 4, so that the first tappet trips and the second trips for the return position.

April 11, 1911   Buttons   989,452

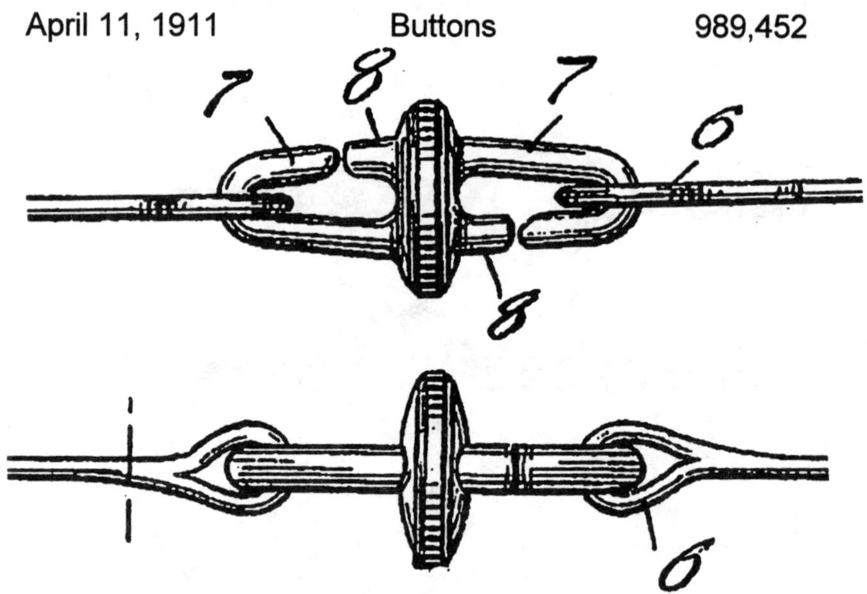

Potosi, Grant County, Wisconsin

"A check-row chain comprising a plurality of links formed of wire, each of said links having an elongated shank portion and eyes at the ends thereof formed by bending and welding, said eyes being throughout of a cross sectional area equal to that of the shank with which they are integrally connected, and obstructing members intermediate said links, and said obstructing members consisting of buttons of malleable metal provided on opposite sides thereof with hooks 7 and with lugs 8, the latter projecting in the direction of the points of the hooks, the hooks of each obstructing member being in engagement with terminal eyes of two links."

The object of the invention is to provide a check-row wire that has no obstruction other than the button, that can catch on planter elements, and a line wire that is strong.

October 22, 1912   Corn Planter   1,041,976

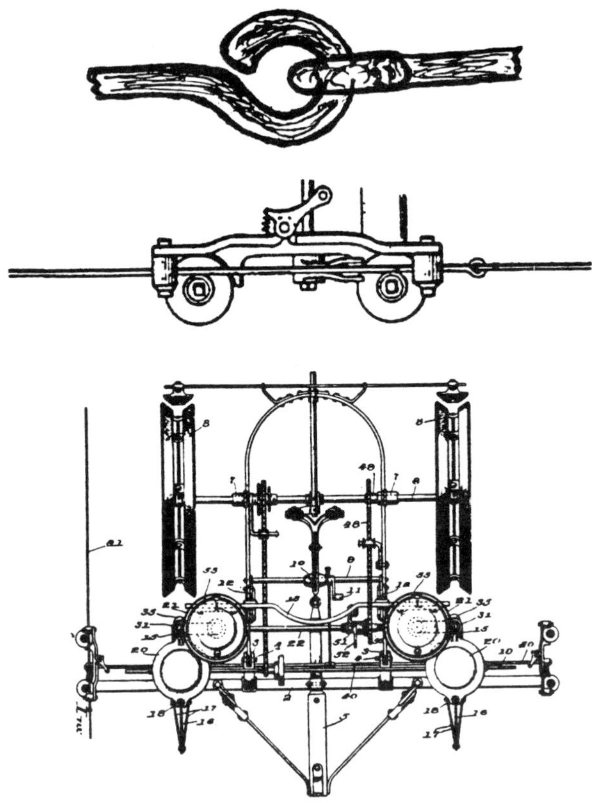

Chicago, Cook County, Illinois

The check-line is barely visible in only one figure. When enlarged it appears to be two inter-locking eyes. The loops or eyes are closed, but not welded. The line wire looks to be heavier than most check-lines, perhaps as large as 12 to 9 gauge.

April 1, 1919   Button   1,298,898

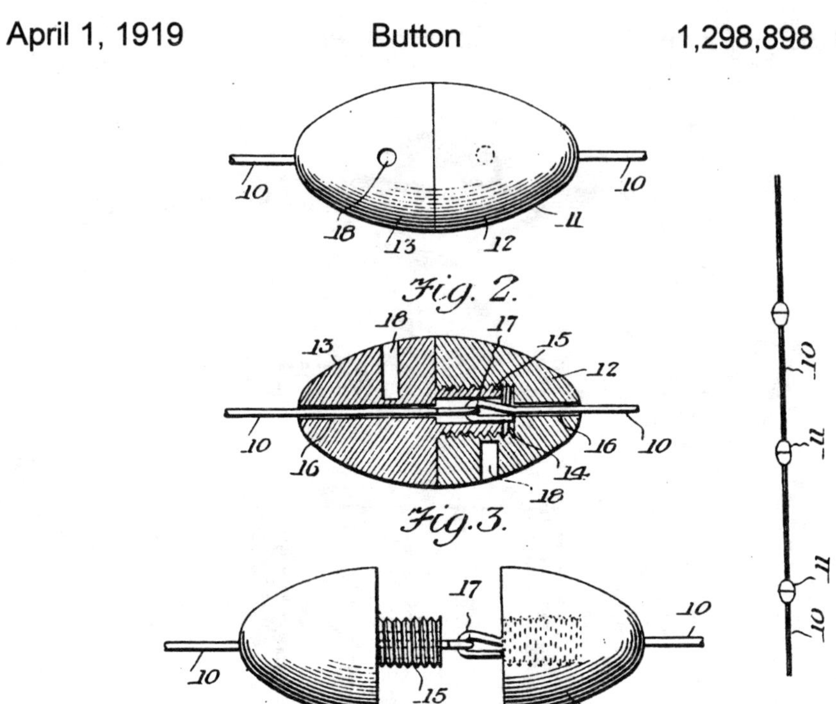

Petersburg, Dinwiddie County, Virginia

"The wire proper is formed of a series of separate sections *10*, each being in length suitable for the distance between buttons *11*, and each button consists of separable members or elements *12* and *13* provided with a threaded socket *14* and stud *15* for engagement."

". . . and when it is desired to pack the wire for transportation or storage, the various sections may be disconnected, or divided up into relatively short sections to form a convenient bundle while retaining the wire element in a straight or extended form to avoid kinking or breaking."

November 27, 1934   Button   1,982,427

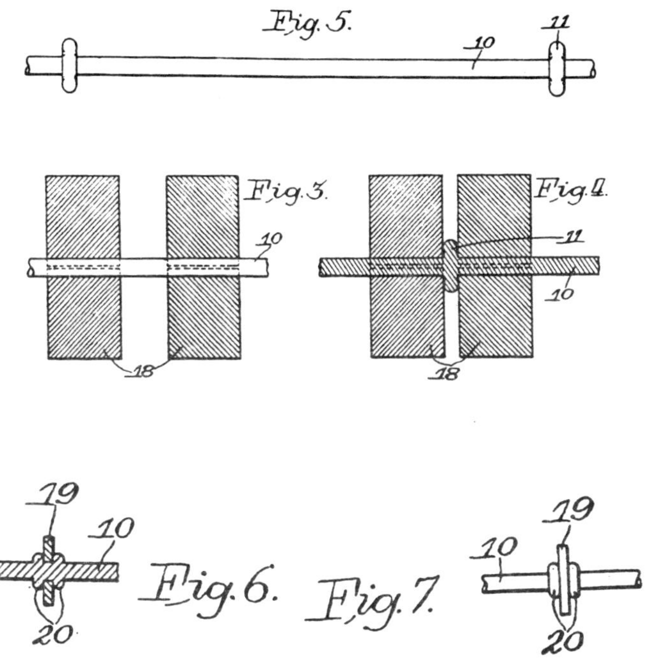

Riverside, Illinois

It is the purpose of this patent to provide a check-row wire continuous in form with an integral button formed from the wire by the use of electrode dies. High carbon wire is used which avoids kinking and stretching such that lower tension is used in planting.

Figs. 6 and 7 show a variation where a hardened washer may be added before upsetting the wire when a more substantial button is wanted. The patent covers both forms.

November 27, 1934   Button   1,982,434

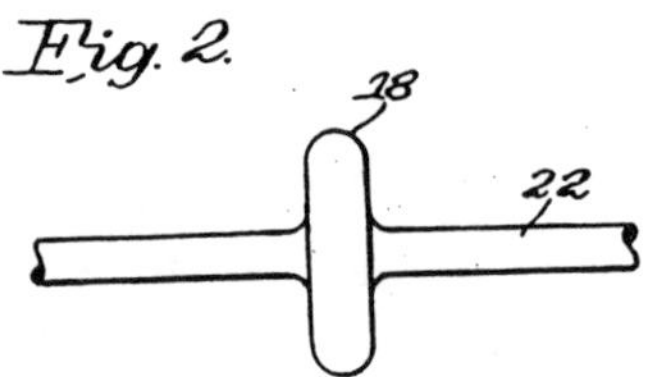

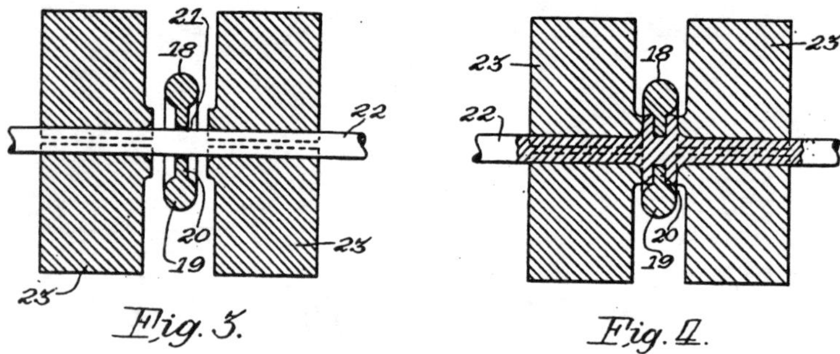

Chicago, Illinois

This patent is very similar to 1,982,427. The difference being a pre-formed disk for the button is used. The upsetting of the wire is less. "... the buttons are preferably formed by employing disk elements *18* formed with torus or doughnut shaped peripheral portions *19* and flat, thin center portions or webs *20*."

This patent was assigned to International Harvester.

SECTION 3

PHOTOGRAPHS
OF
COLLECTION SPECIMENS

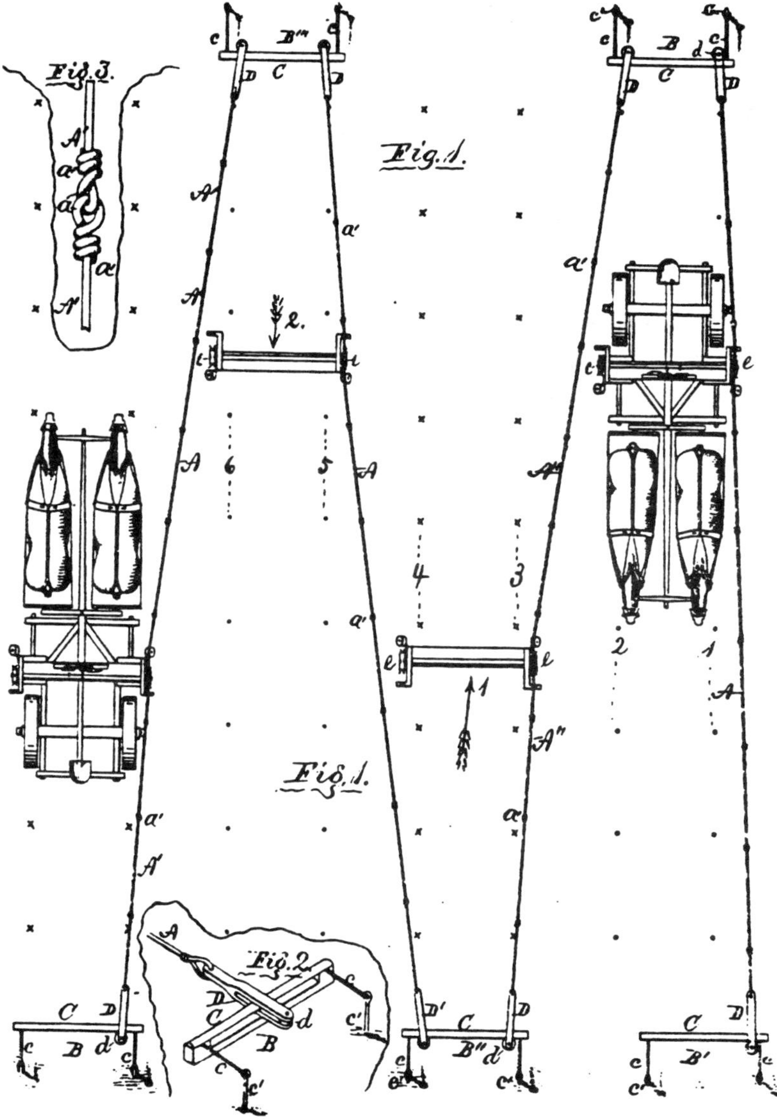

HAWORTH PLANTER WIRE REEL

George and Lysander Haworth Reel used originally for 119,142 and 159,177 rope check-lines. Roof of wooden reel has the Haworth names and patent dates September 19, 1871 and January 26, 1875.

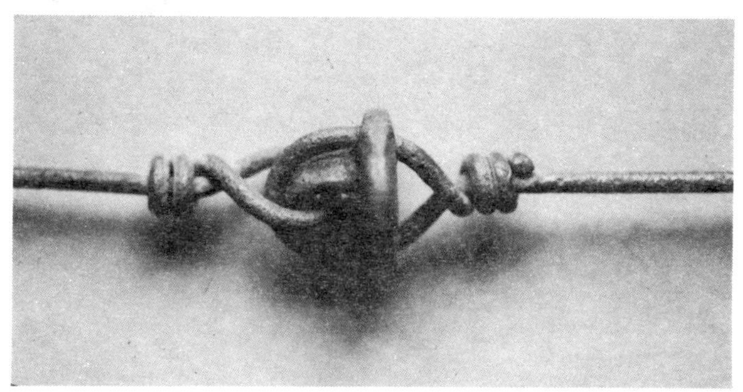

A. Phelps (250,750) Variation

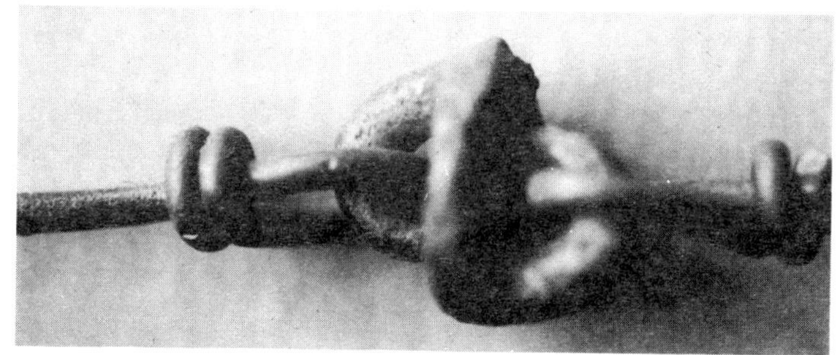

B. Phelps (250,750) Variation

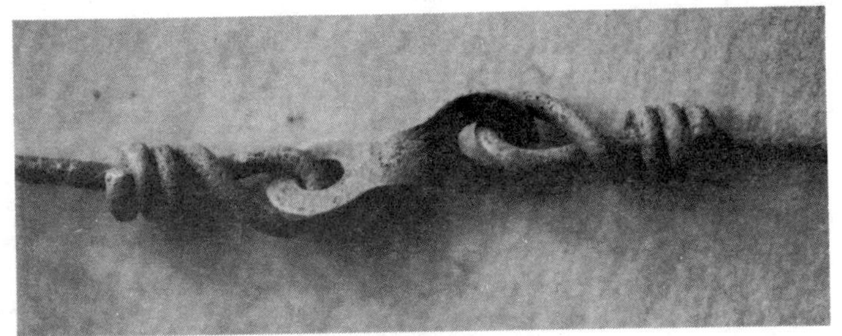

A. Callan (274,266) Variation

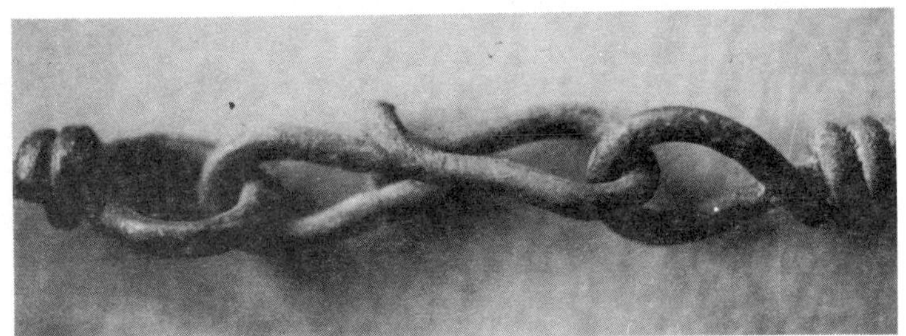

B. Flynn (442,032) Variation

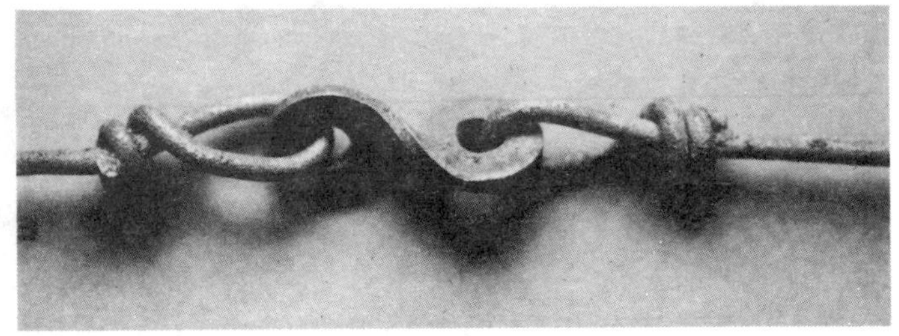

C. Callan (274,266) Variation

A. Unknown

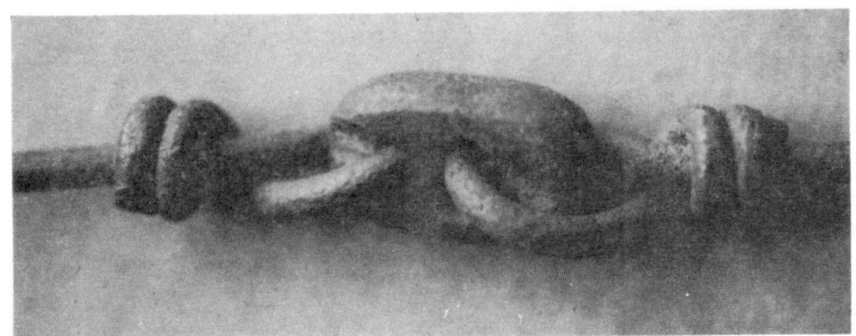

B. Unknown

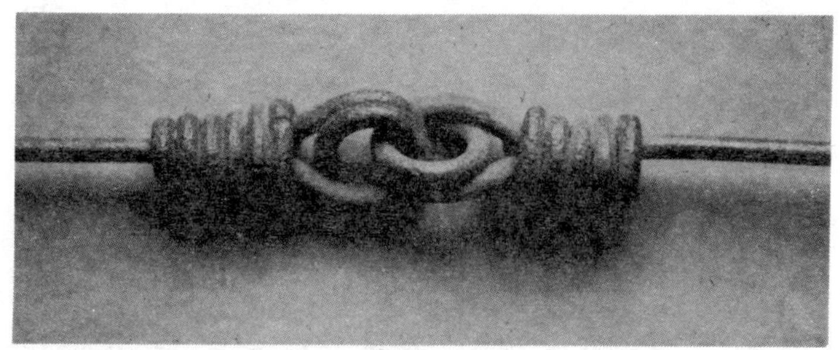

A. Parker (628,959) Variation

B. Parker (628,959) Variation

C. Parker (628,959) Variation

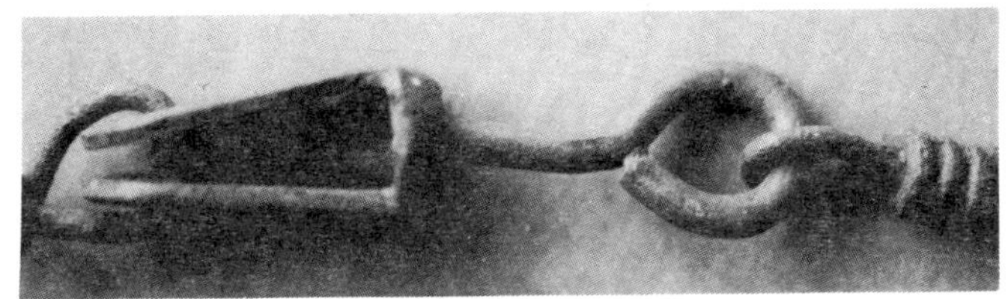

A. Unknown

B. Unknown
Origin possibly is the maker of the net tie stay clip.

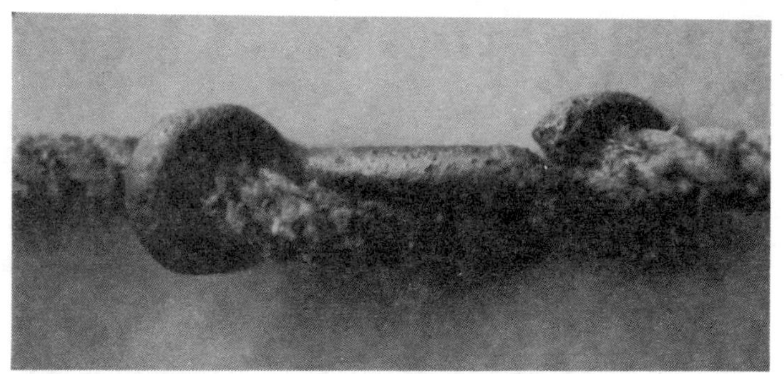

A. Feldmier (255,962) Variation

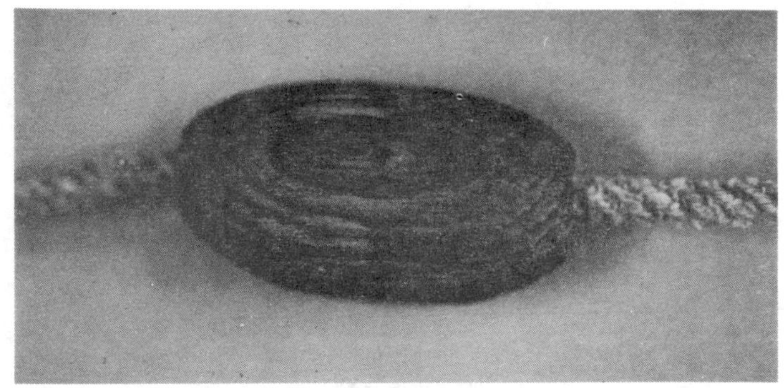

B. Cross (210,998) Variation

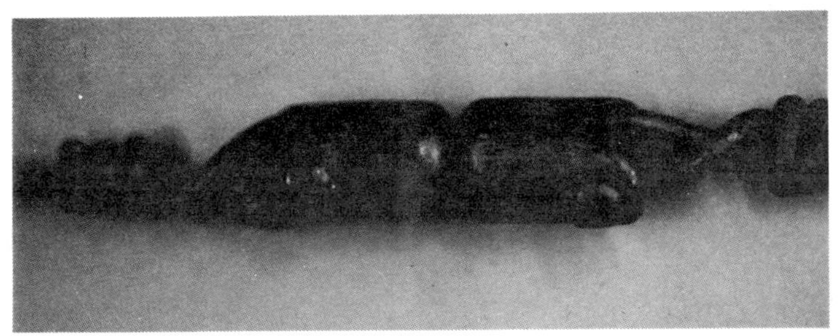

A. Unknown Coupler

B. Unknown Splice

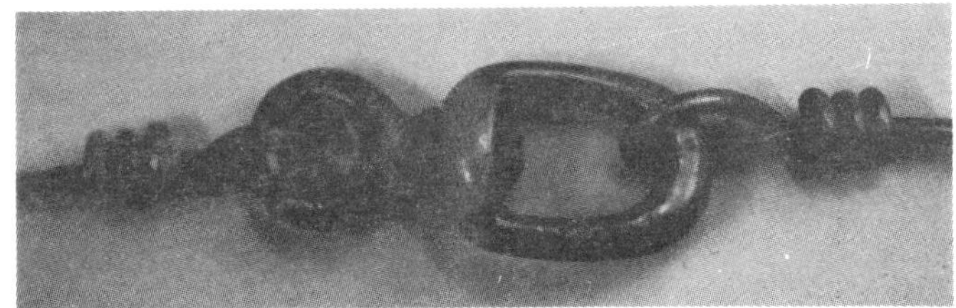

C. Unknown Swivel

A. Unknown Coupler

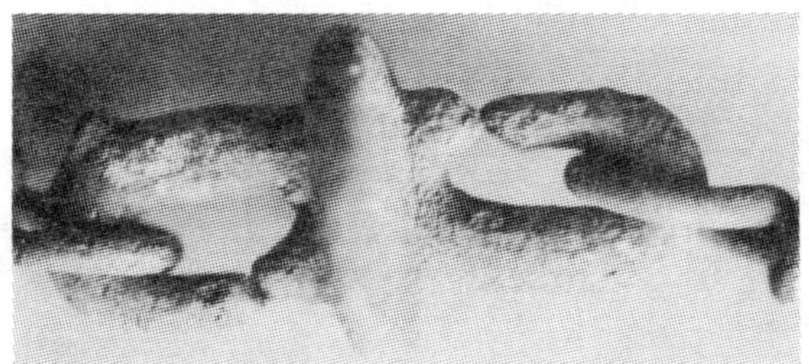

B. Van Natta (989,452) Variation

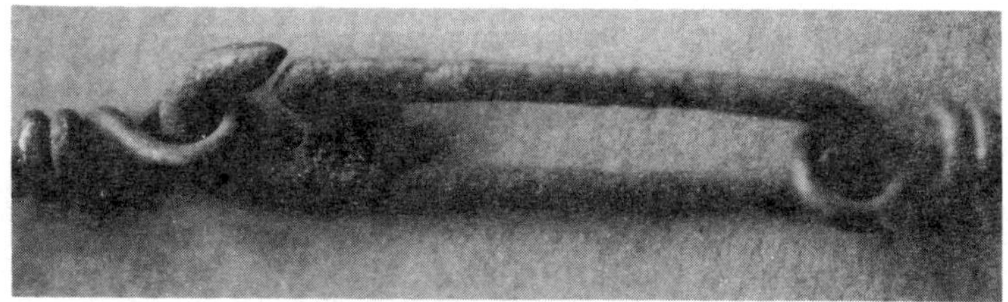

C. Unknown Coupler

A. Hyer (234,780) Variation

B. Hyer (234,780) Variation

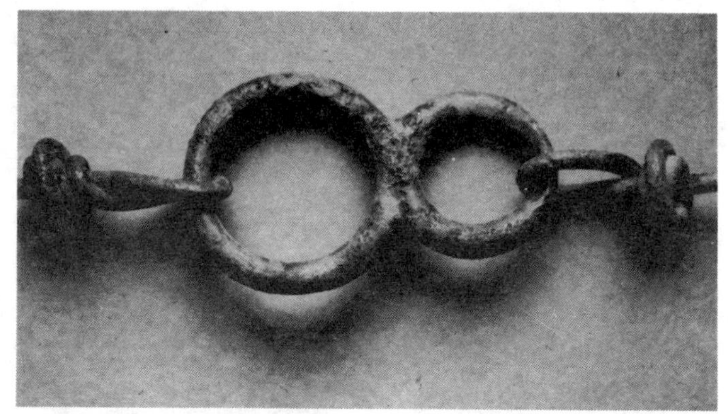

A. Baxter (866,708) Variation

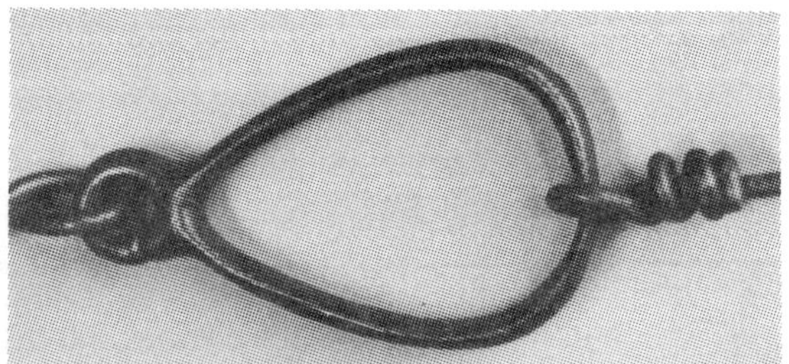

B. Baxter (866,708) Variation

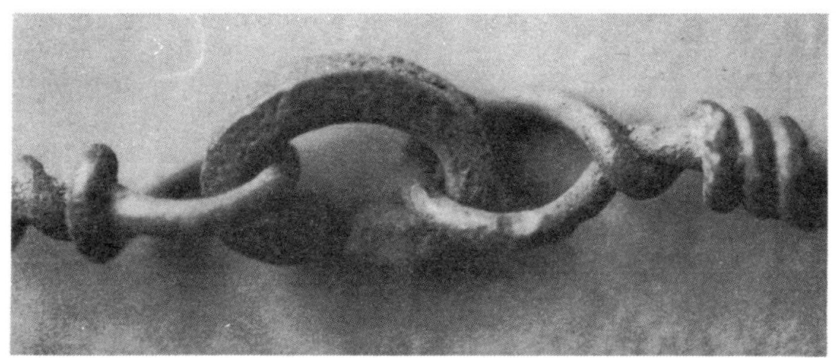

A. Hyer (234,780) Variation

B. Barnes (132,792) Splice

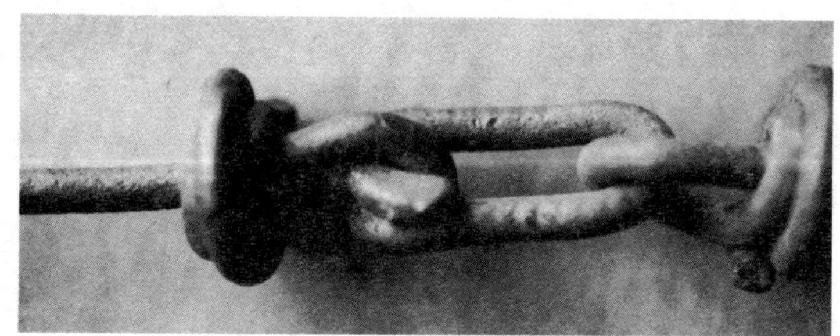

A. Barlow (328,452) Splice

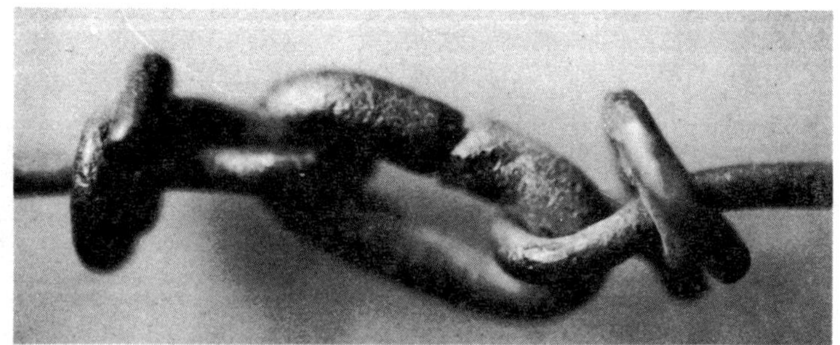

B. Barlow (328,452) Splice

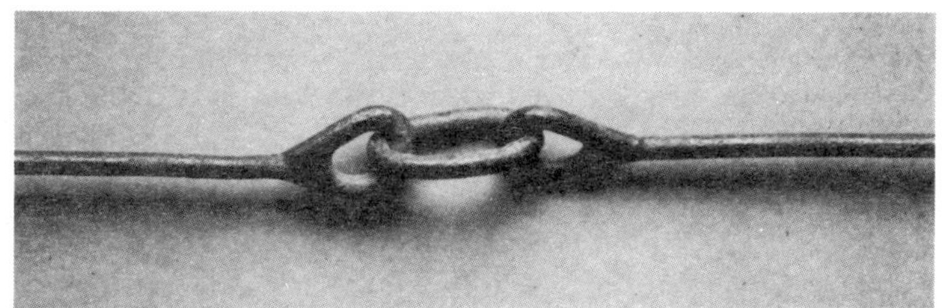

A. Caviness (234,243) Variation

B. Unknown Crimped brass

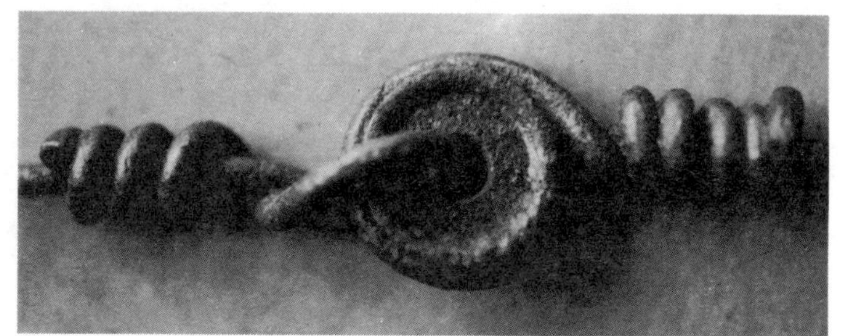

A. Unknown

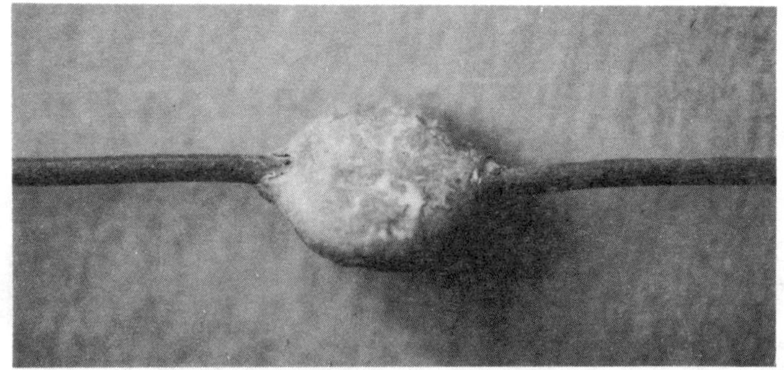

B. Locke (296,981) Variation

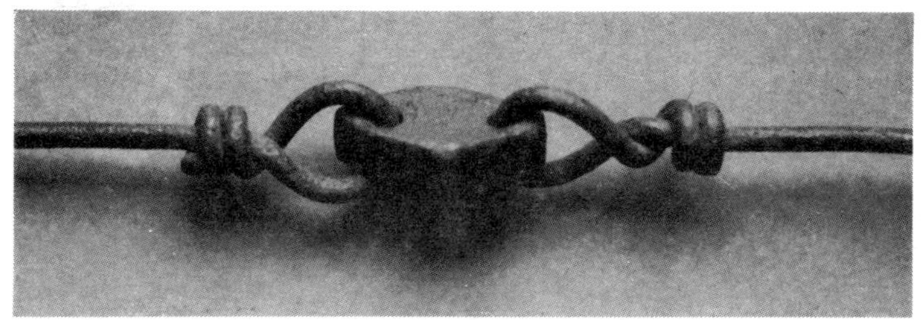

A. Unknown

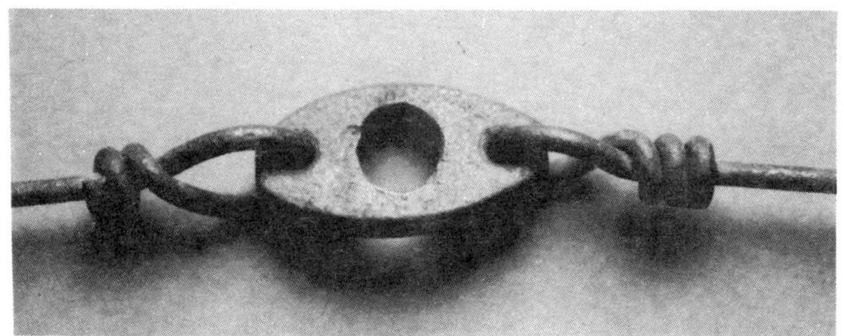

B. Unknown

C. Unknown

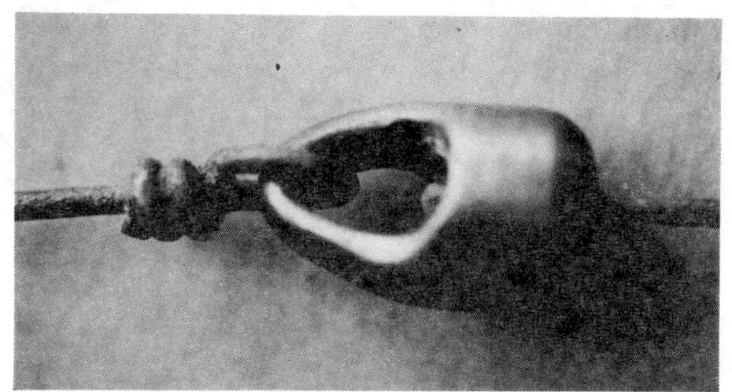

A. Smith (360,825) Brass

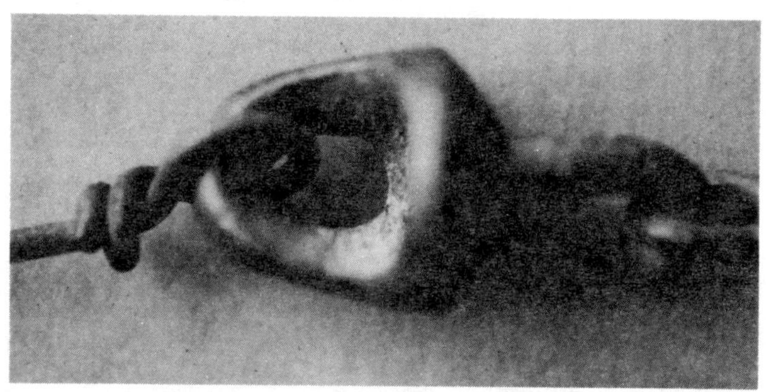

B. Smith (360,825) Variation

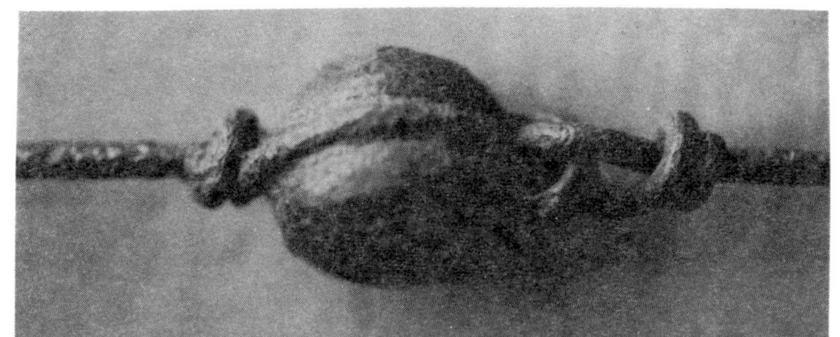

A. Thomson (289,468) Variation
Commonly called the "Padlock"

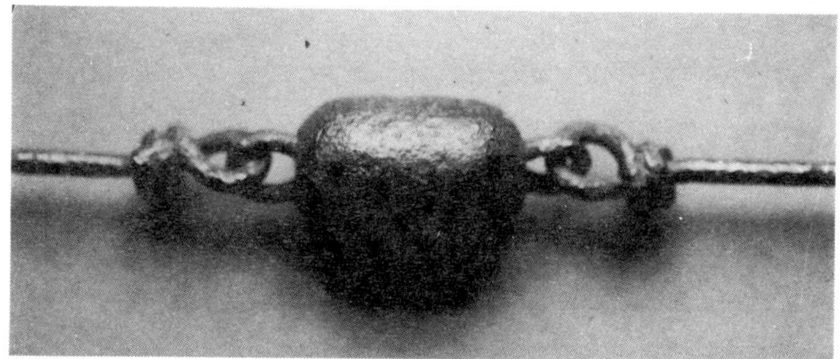

B. Iles (222,278) Cast Knot

A. Unknown Brass Ring

B. Unknown

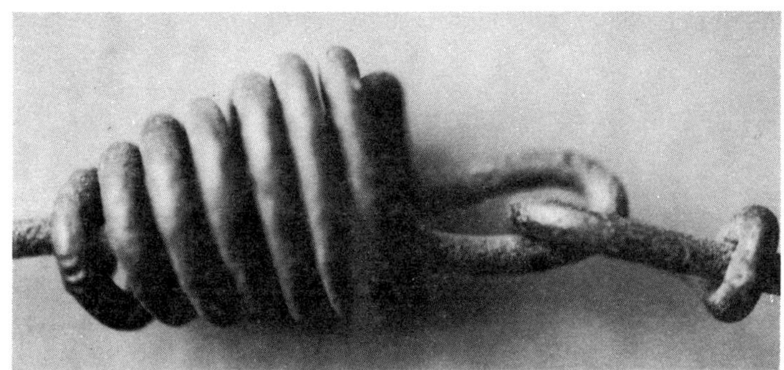

A. Parker (578,830) Commonly called the "Bee-Hive" Knot

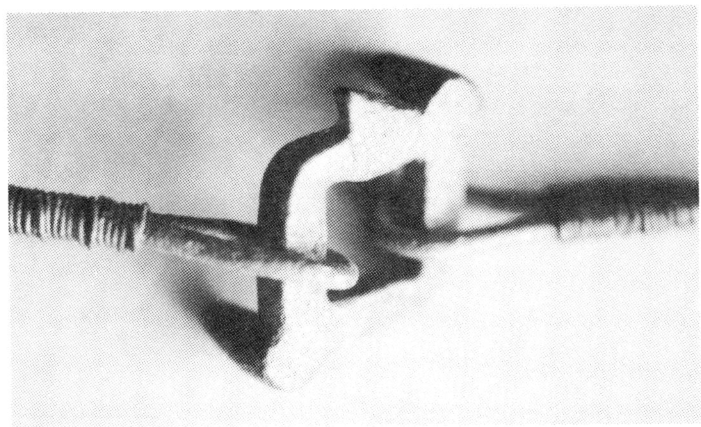

B. Unknown
Possibly a splice or marker in a hand-made check-line used for hand planting. The knot is made from a cast-iron square shaft dog from some implement.

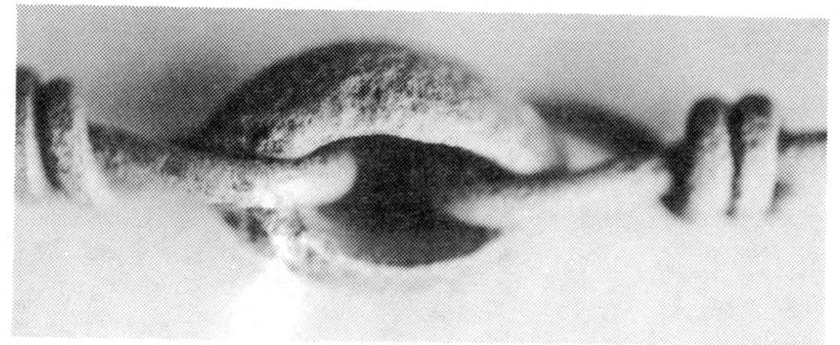

A. Unknown
Commonly called "Football"

B. Hyer (234,780) Variation

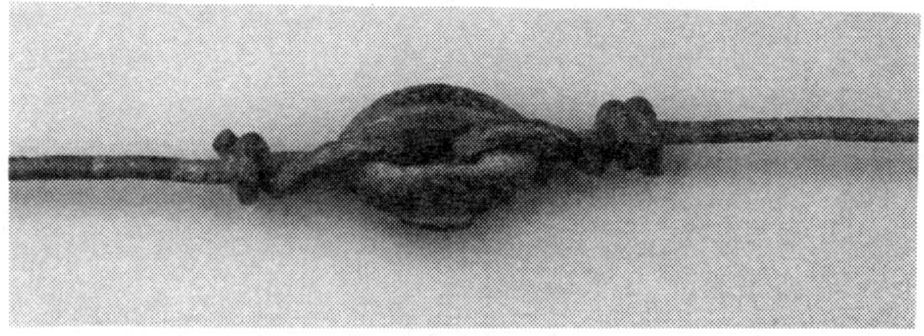

C. Unknown
Commonly called "Football" with a rectangular slot.

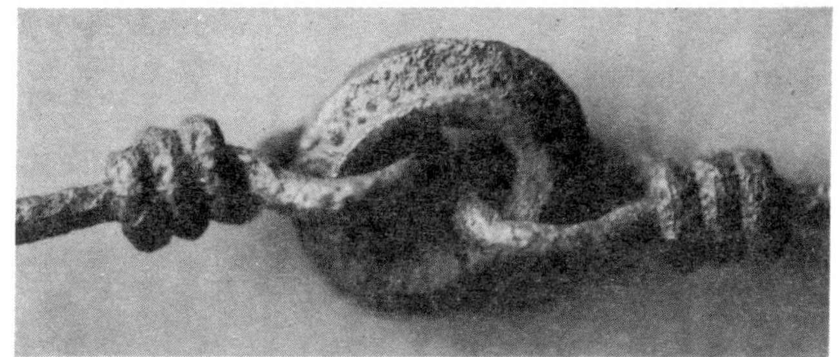

A. Hyer (234,780) Variation

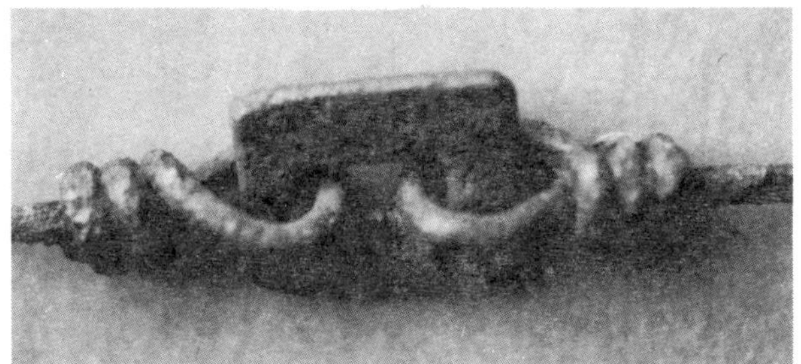

B. Hyer (234,780) Variation

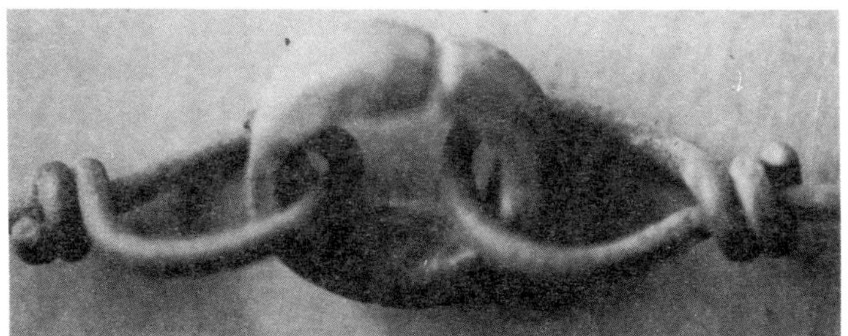

C. Hyer (234,780) Variation

A. Grush & Lockhart (233,927) Holed Rectangle Variation

B. Grush & Lockhart (233,927) Square Block Variation

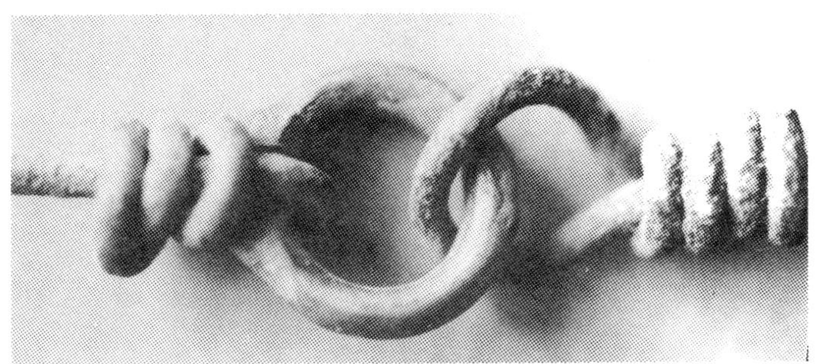

A. Avery (537,261) Brass Knot Variation

B. Avery (537,261) Brass Knot Variation

A. Possibly an anchor link that could have been used as a visual marker when hand planting.

SECTION 4

SOME FOUND KNOTS

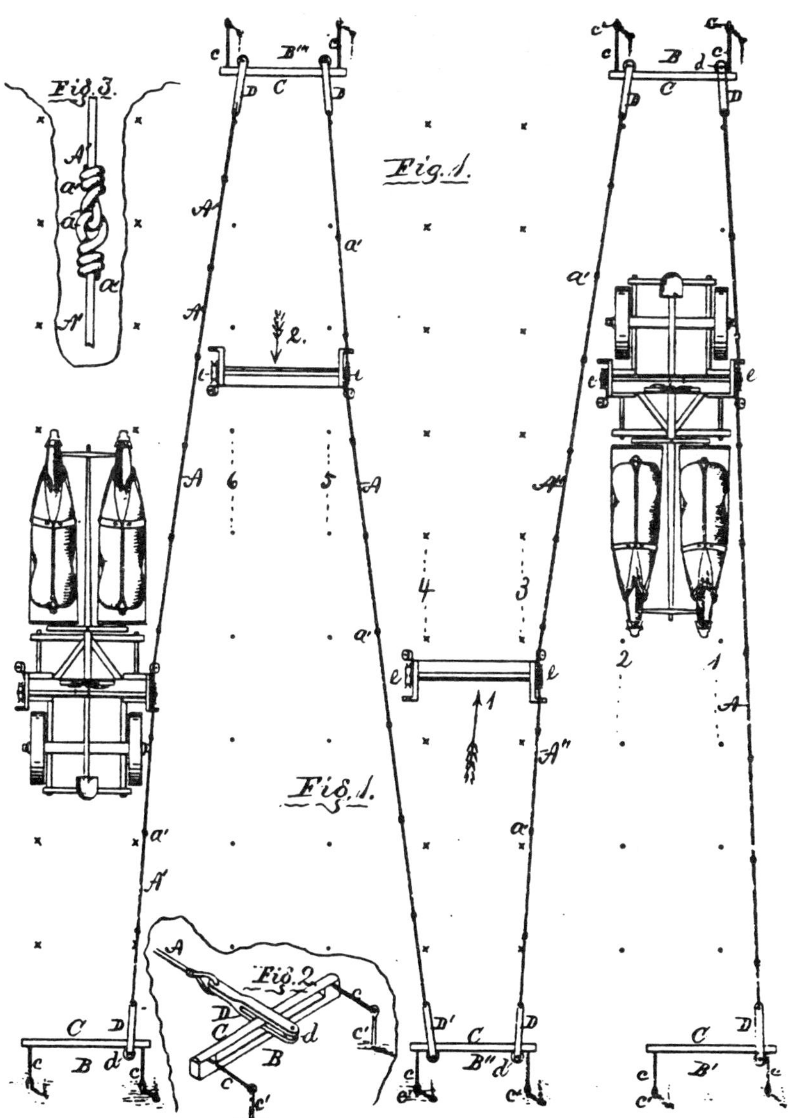

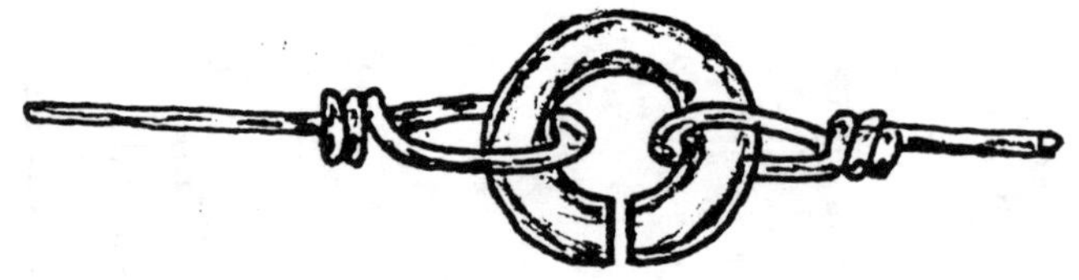

A. Hyer (234,780) Variation

B. Hyer (234,780) Variation

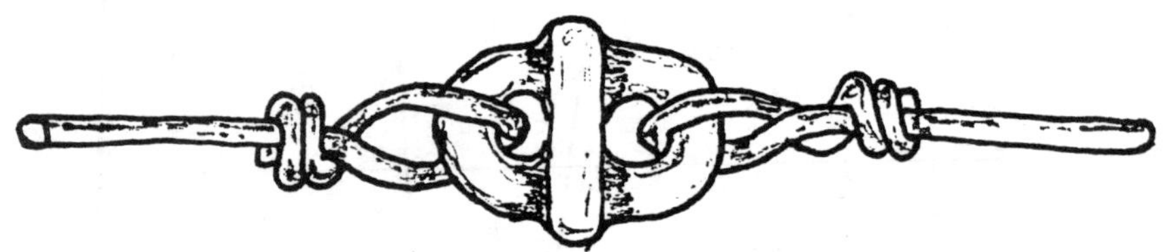

C. Phelps (250,750) Variation

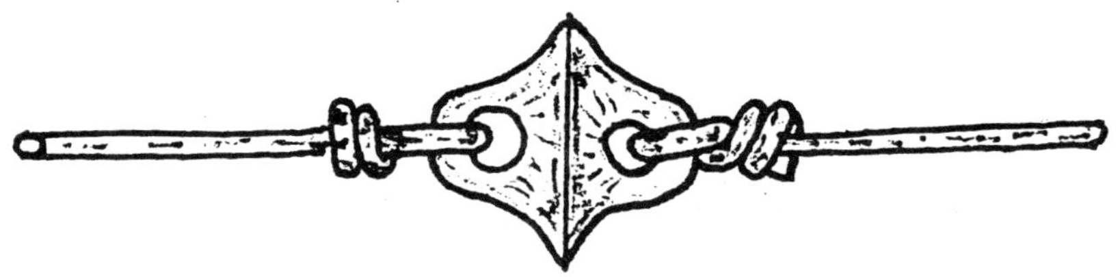

A. Phelps (250,750) Variation

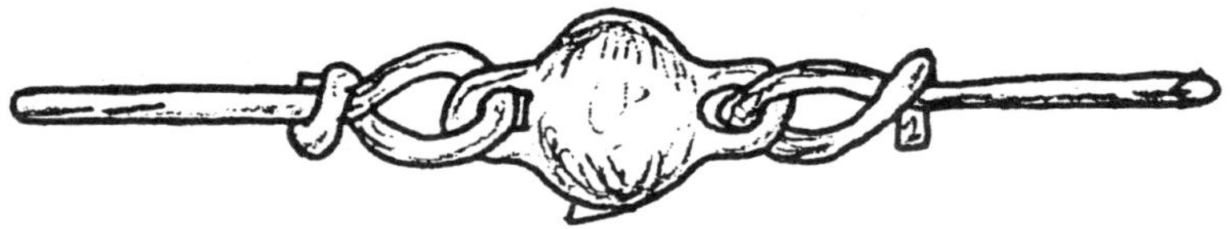

B. Thomson (289,468) Variation

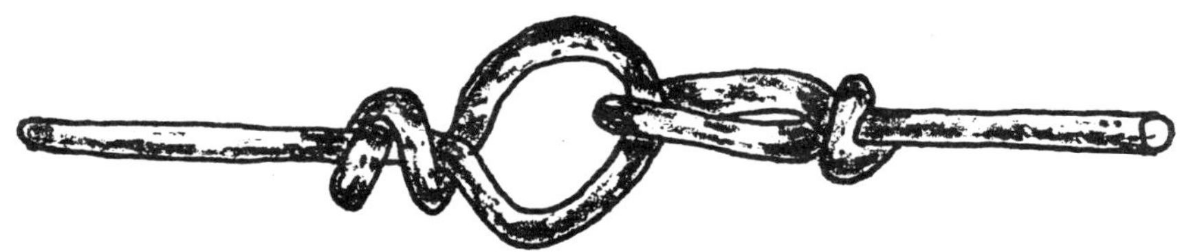

C. Barnes (132,792) Variation

A. Unknown

B. Barlow (328,452) Variation

C. Barnes (132,792) Variation

A. Phelps (250,750) Variation

B. Avery (537,261) Hog Ring Splice

C. Phelps (250,750) Variation

A. Barlow (328,452)
Cast Splice

B. Barlow (328,452)
Cast Splice

A. Unknown

B. Unknown

C. Lord (232,137) Brass Variation

ANCHORS AND REELS

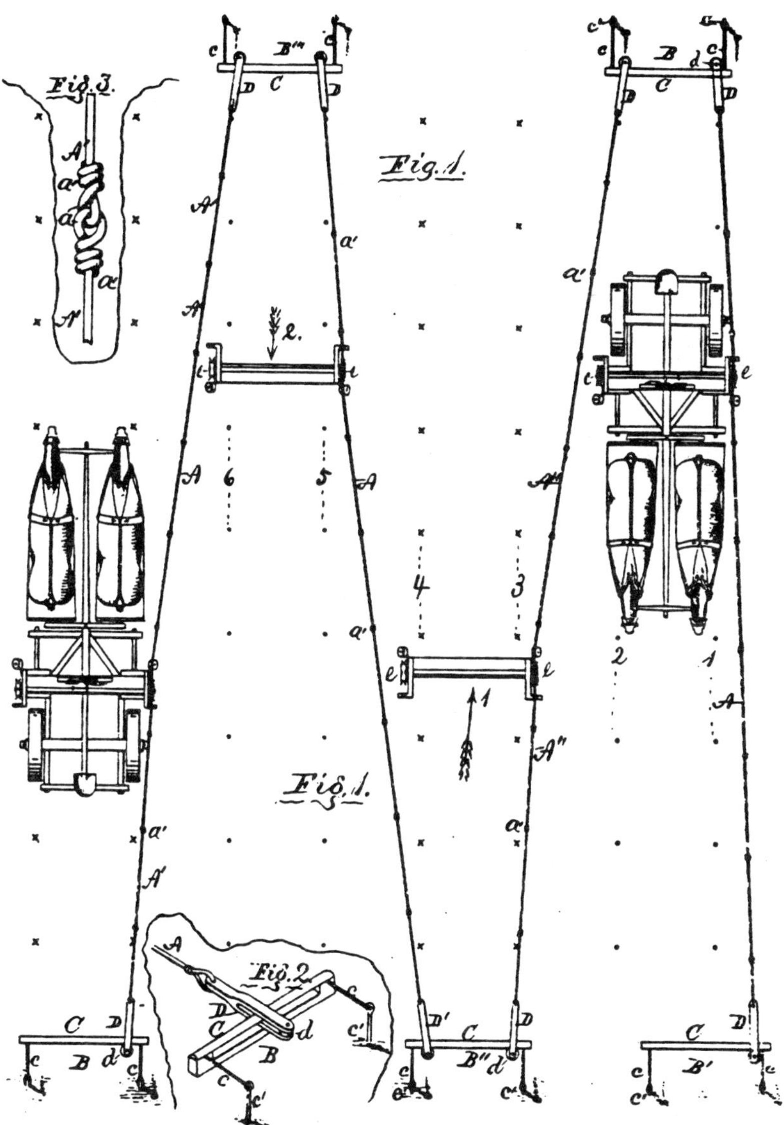

March 23, 1875   Tension Anchor   161,232

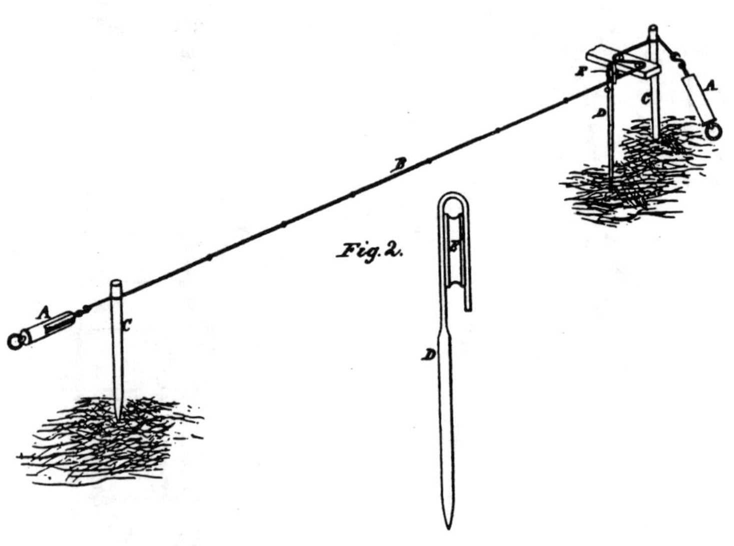

Atlanta, Logan County Illinois

"In the use of the common check-rower it requires a good deal of judgement as to the amount of force to be applied in tightening the rope, and, as many persons using it lack the necessary judgement as to the amount of force to be applied in case of ravines in the field, or in case of irregular tracts of ground to be planted, the common result is irregular rows one way of the field."

"This defect I have successfully overcome by the simple use of two ordinary spring scales. The hills will then correspond exactly, and the rows will be regular both ways of the field."

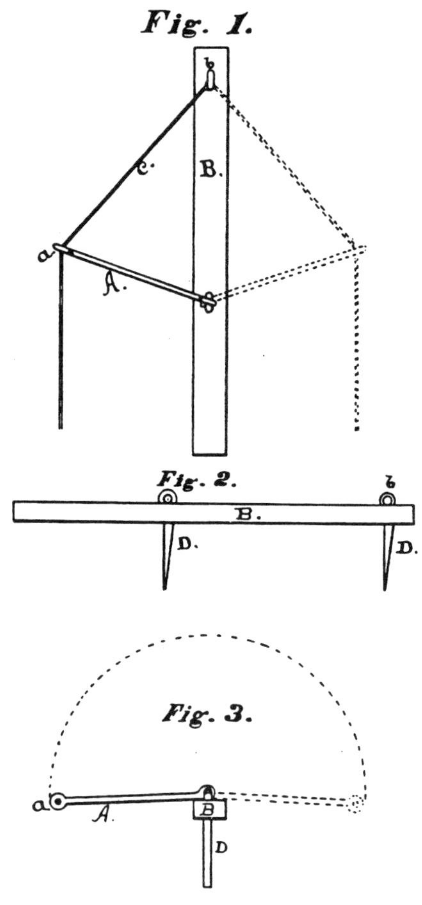

Decatur, Macon County, Illinois

November 19, 1879   Anchor   210,109

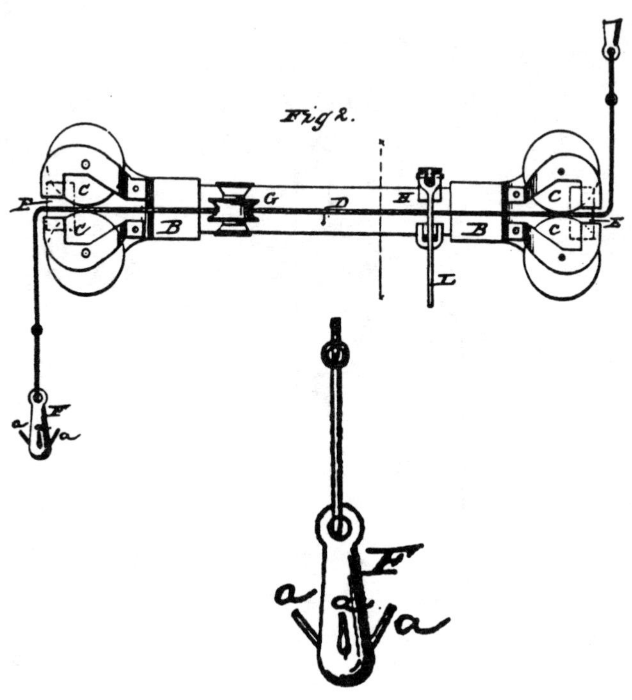

Crescent City, Iroquois County, Illinois

"My invention is intended as an improvement upon what is known as the 'Haworth Check-Rower'; and it consists in the arrangement of pulleys at each end of the beam, and in a clamping device for holding the rope while turning the planter, all as hereinafter more fully set forth."

The significant item from this patent related to the rope is the weight anchor per Fig. 2. It resembles a boat anchor and is intended to keep the rope taut while turning.

December 9, 1879   Tension Anchor   222,421

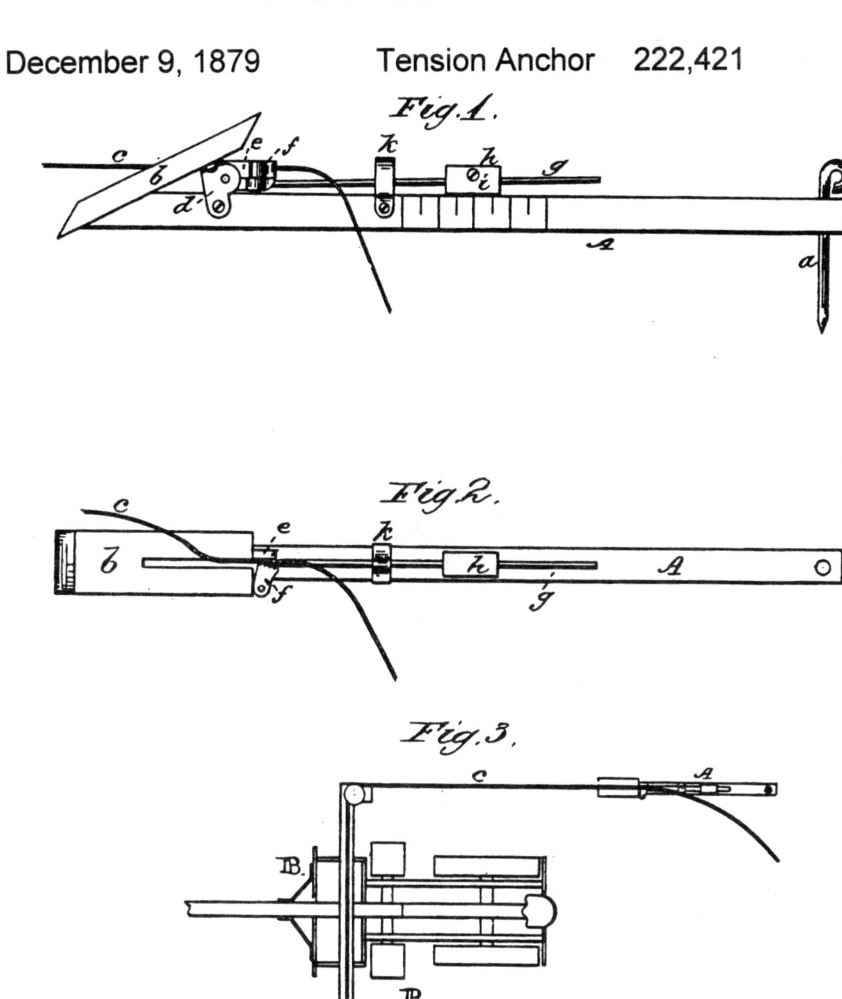

Tolono, Champaign County, Illinois

"The tension of the rope may be regulated by moving the weight h along the rod g any required distance, the scale on the side of the stake indicating the amount of adjustment necessary to suit various kinds of planters."

December 23, 1879   Reel   222,854

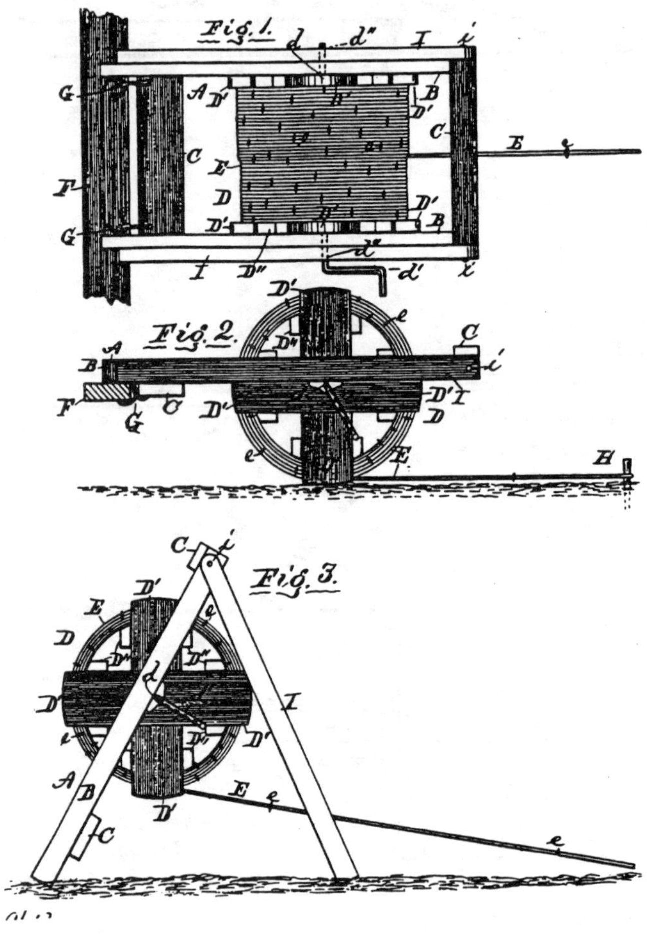

Galesburg, Knox County, Illinois

GEORGE D. HAWORTH

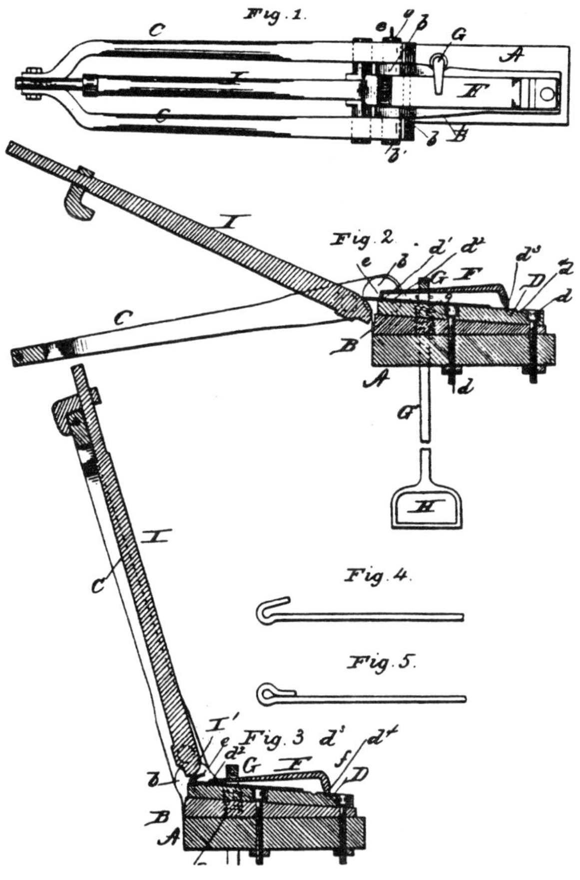

Decatur, Macon County, Illinois

"My invention has especial reference to the looping or knotting of the ends of the links or sections of wires for the use in connection with check-row corn planters; but it may be advantageously employed for forming loops in wires for other uses, ..."

December 28, 1880   Reel   236,024

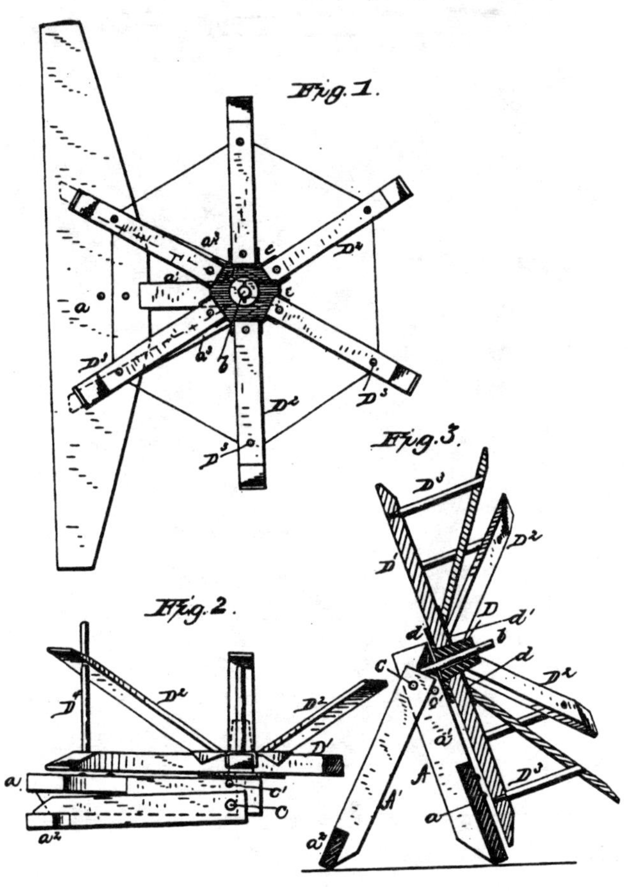

Decatur, Macon County, Illinois

January 11, 1881   Anchor   236,580

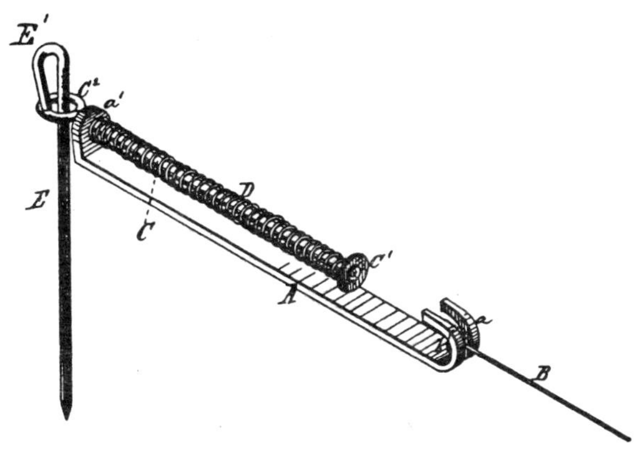

Decatur, Macon County, Illinois

January 11, 1881   Anchor   236,581

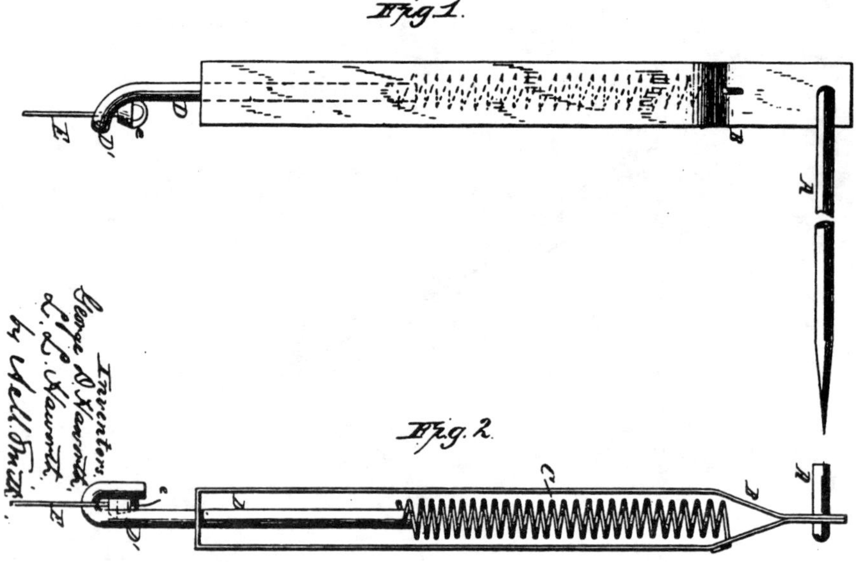

Decatur, Macon County, Illinois

NATHAN C. BOLIN

July 28, 1882   Laydown/Takeup Machine   261,658
For Check-Row and Barbed Wire

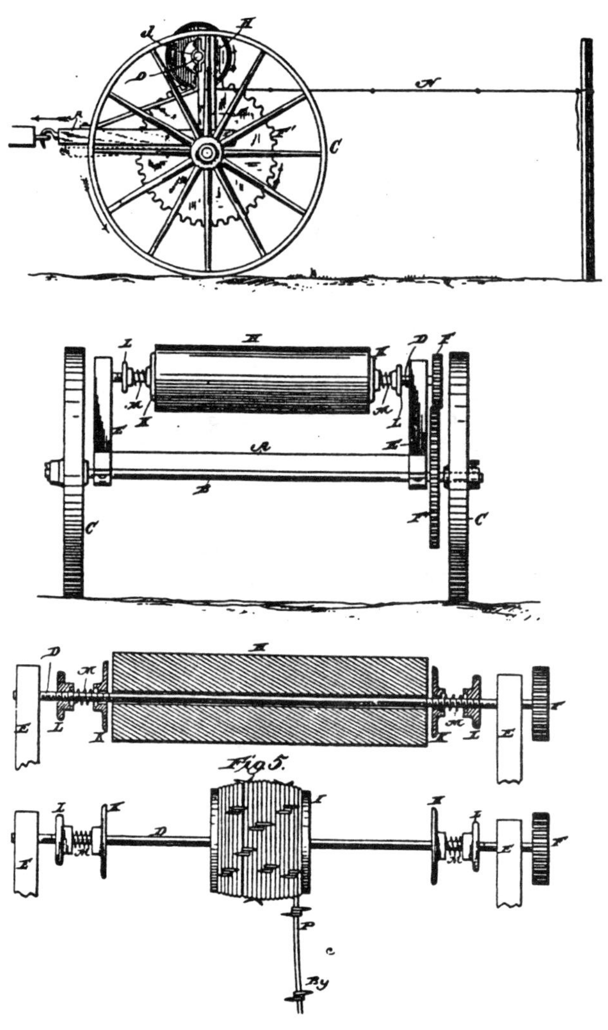

Red Oak, Montgomery County, Illinois

September 26, 1882   Reel   265,197

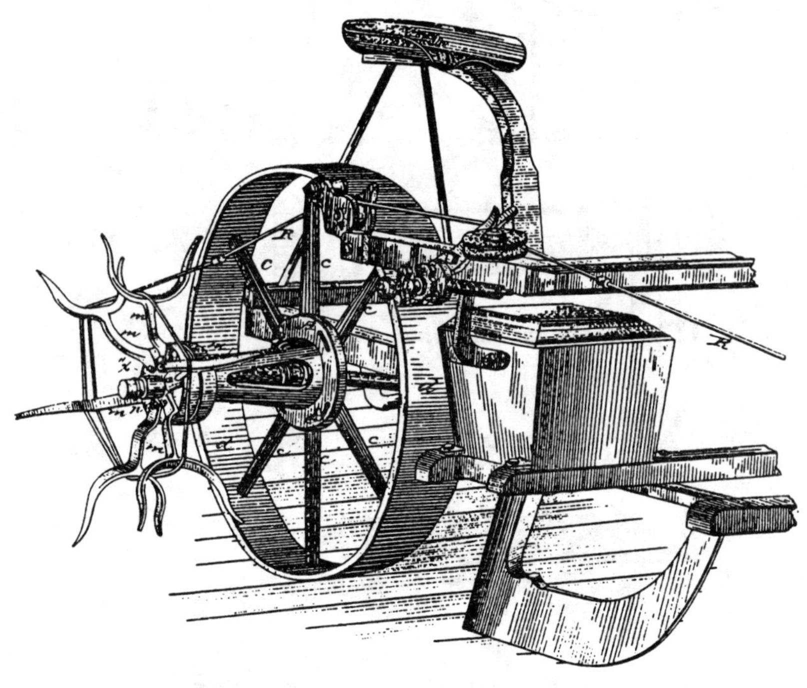

Mason City, Mason County, Illinois

October 17, 1882   Shift Anchor   266,250

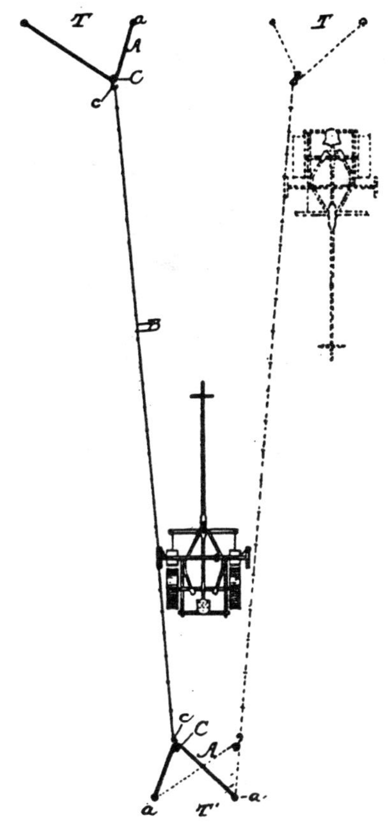

Decatur, Macon County, Illinois

"$\mathcal{A}$ designates the main portion of the anchor, which is formed of any suitable flexible material, preferably cable-wire, wire strand, chain or rope. The stakes pass through rings at both ends."

TYLER C. LORD

November 28, 1882   Anchor   268,110

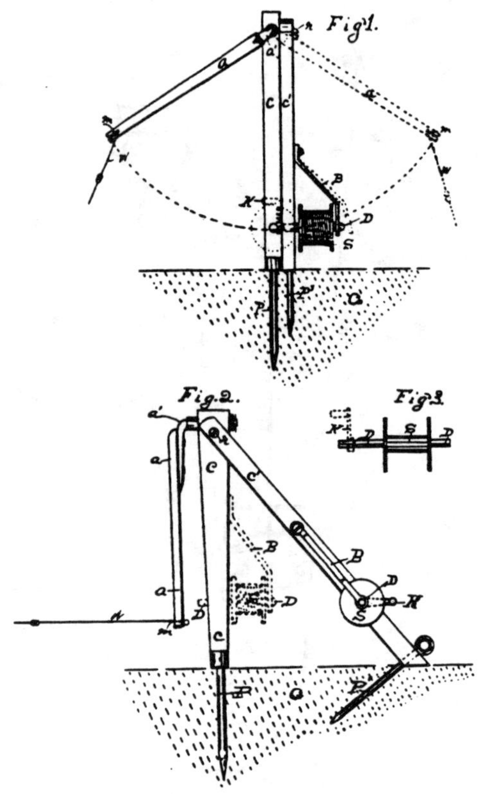

Joliet, Will County, Illinois

"The nature of this invention is for the purpose of holding the cord or wire of a check-rower at either end of the field in such a manner that the anchor may move or shift the cord or wire to accommodate it to the position of the check-rower on its journey across the field and back."

February 20, 1883   Flexible-Anchor   272,402

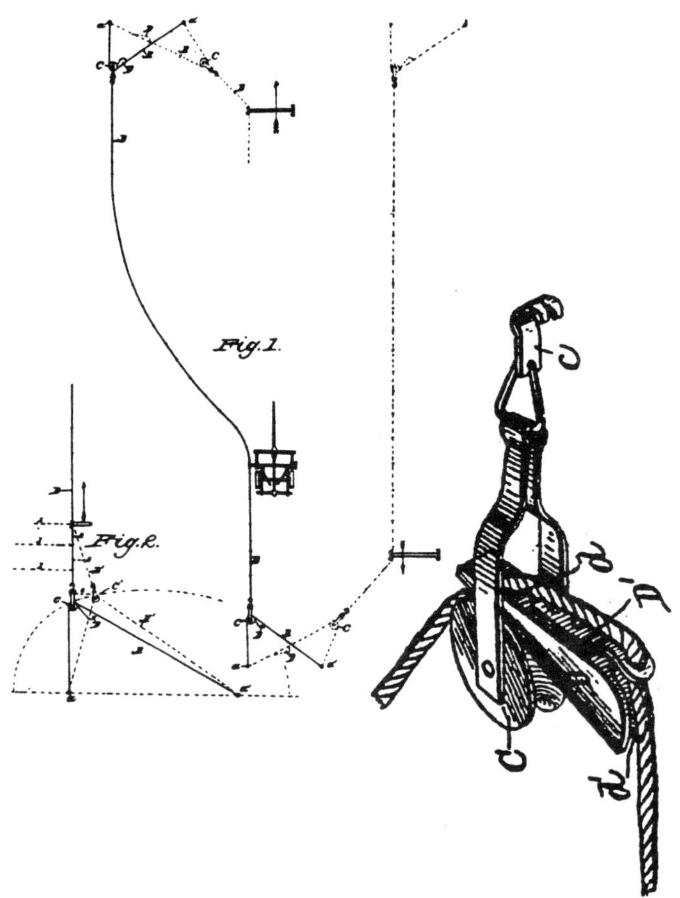

Bloomington, McLean County, Illinois

"My invention relates to anchors for check-row lines, which serve by a lateral shift to properly position the check-line during a double traverse of the field by the corn planter."

A flexible anchor is described.

March 6, 1883   Anchor   273,263

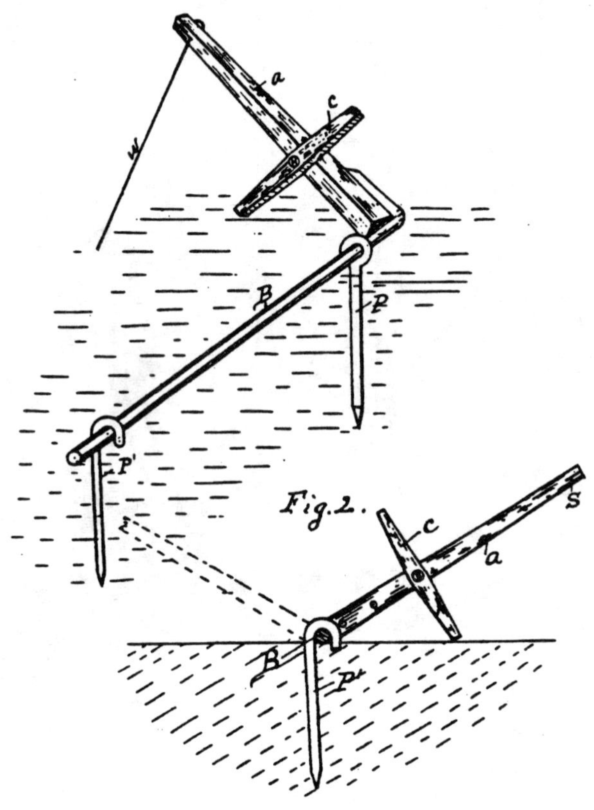

Joliet, Will County, Illinois

"This invention relates to a device for the purpose of holding or anchoring a check-row wire at either end of a field, and is of that class that rotate to carry the wire forward to be in line with the planter as it travels across the field."

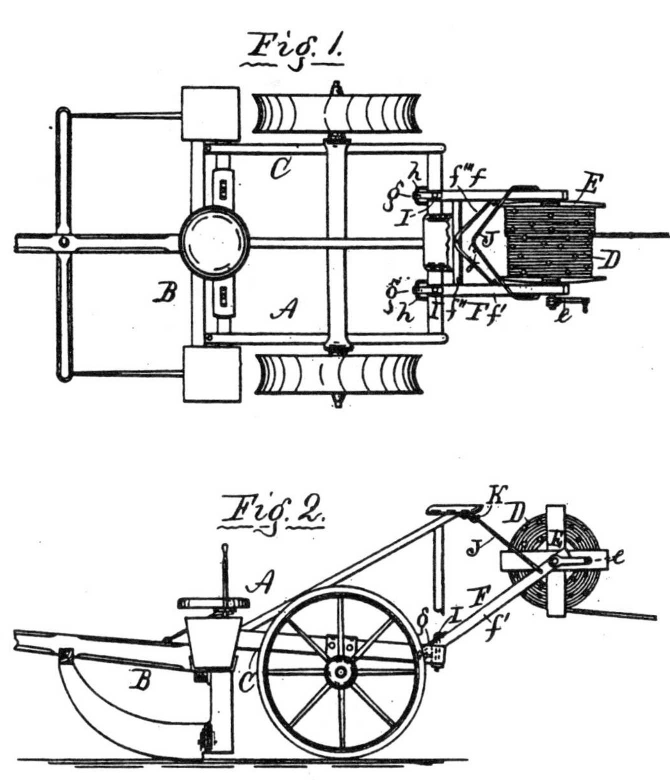

Alpha, Henry County, Illinois

May 8, 1883   Reel   277,030

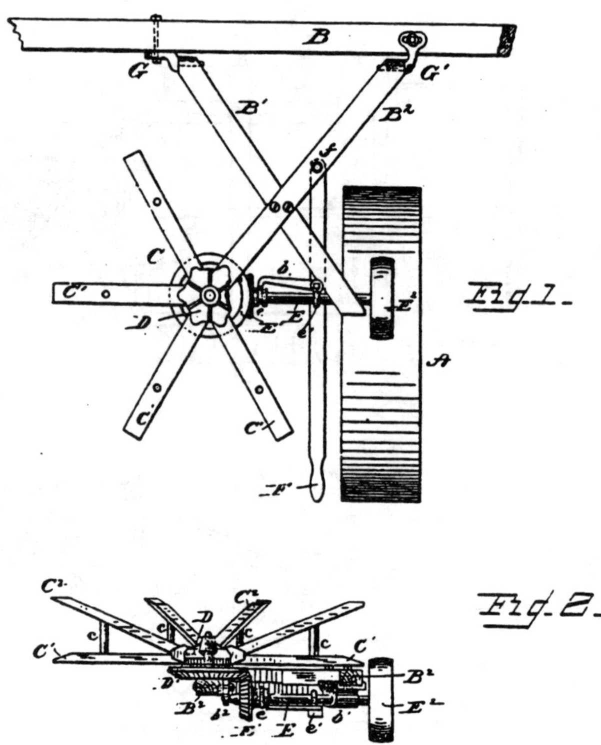

Decatur, Macon County, Illinois

GEORGE W. BROWN AND JOHN C. TUNNICLIFF

May 29, 1883 Laydown/Takeup Machine   278,653

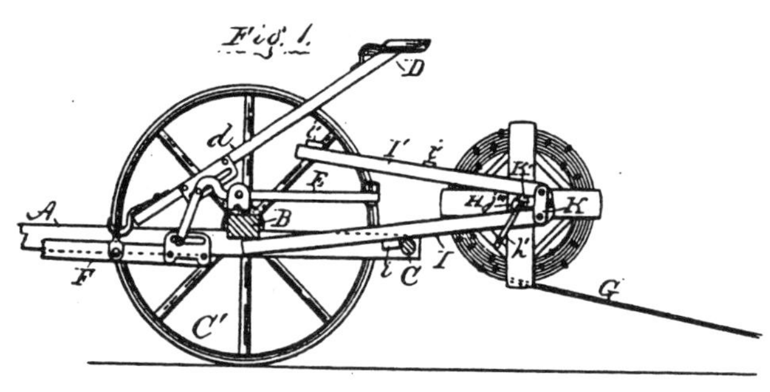

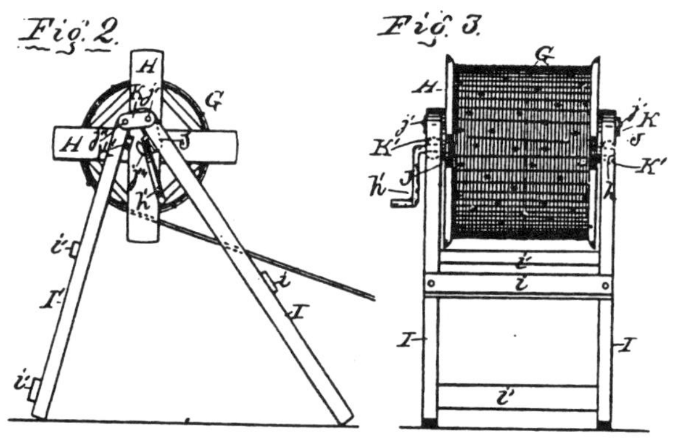

Galesburg, Knox County, Illinois

December 4, 1883   Anchor   289,372

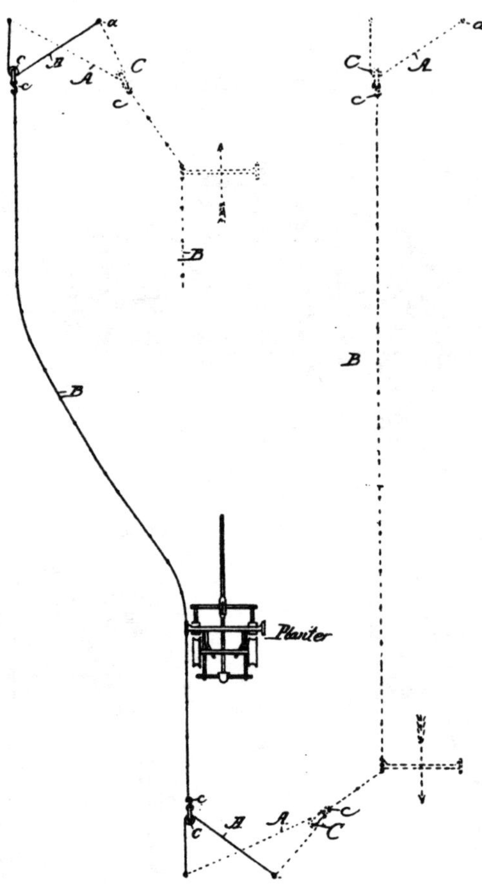

Bloomington, McLean County, Illinois

"My invention relates to anchors sustaining check-row wires which will permit of a lateral shift or adjustment of the check-line. It consists in the form of a flexible cable, chain, or rope provided with suitable connections."

May 13, 1884   Reel   298,678

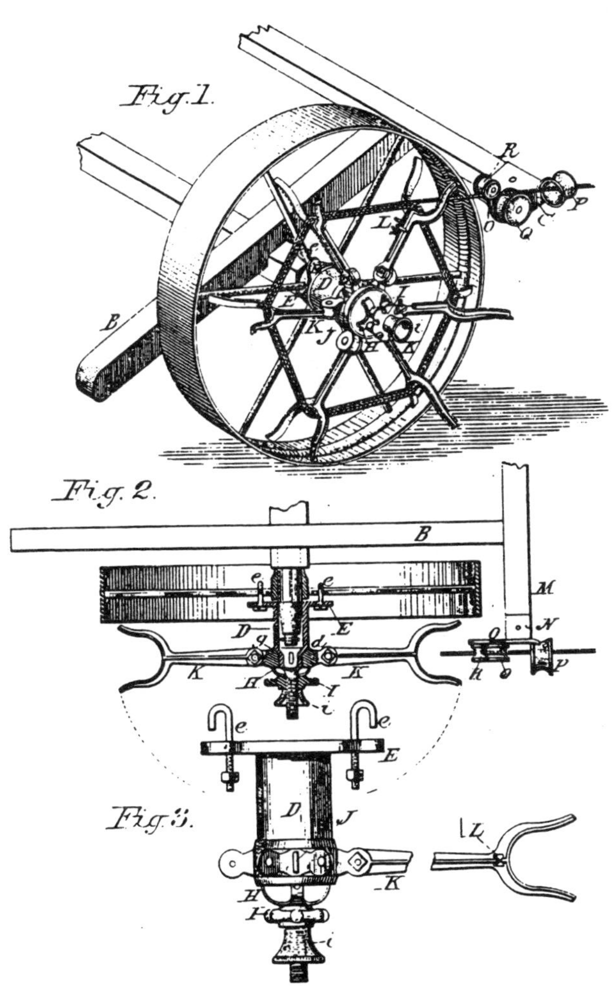

Decatur and Mason City, Illinois

January 6, 1885　　　　Anchor　　　　310,384

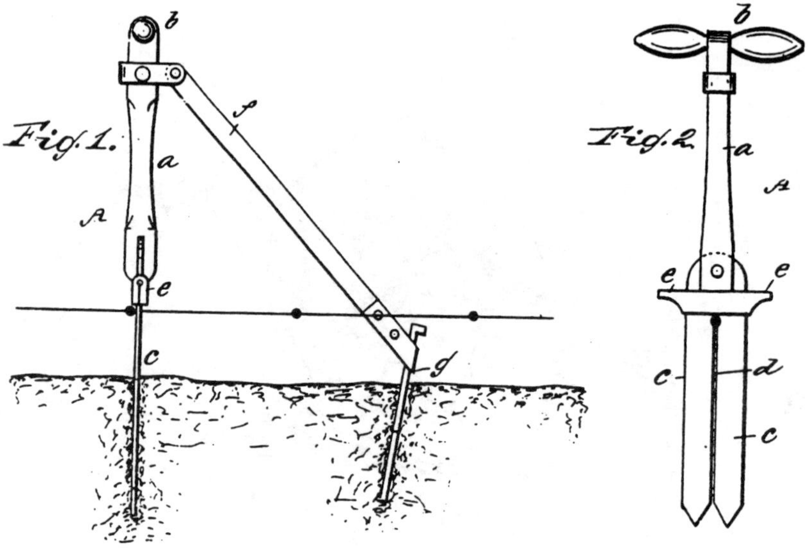

Fig. 1.

Fig. 2.

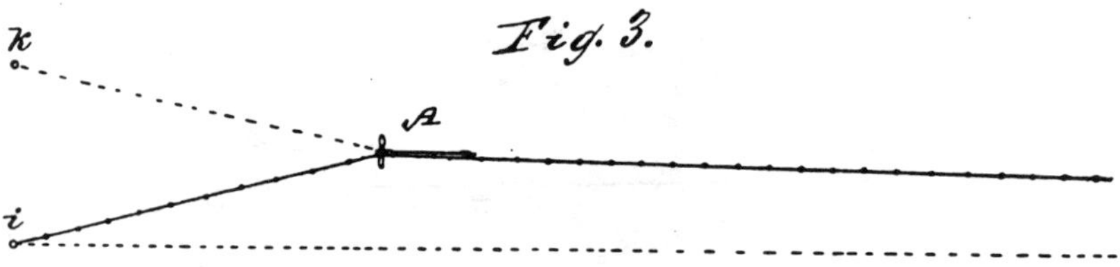

Fig. 3.

Paris, Bourbon County, Kentucky

"My invention allows for resetting stakes without a change in tension. A stake holds the wire before the wire is moved."

ABIJAH T. CROW

October 20, 1885   Equalizer Anchor   328,879

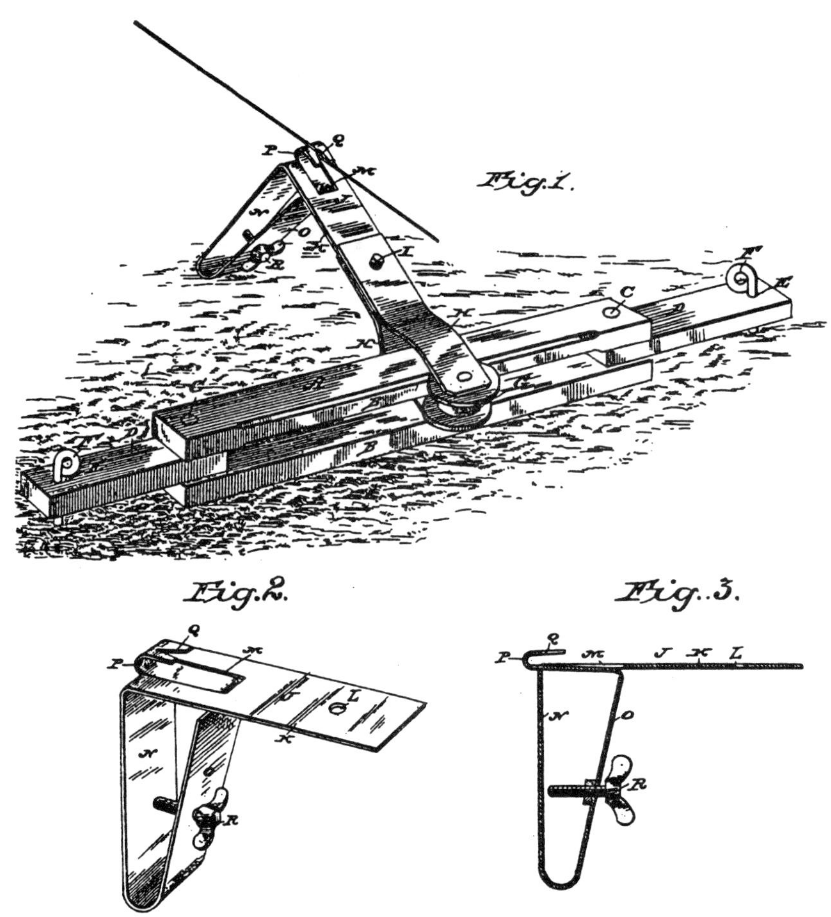

Clarinda, Page County, Iowa

"My invention relates to anchors or equalizers for the tappet wires of check-row corn planters and shall be so constructed as to retain the wires at all times at an equal degree of tension."

October 27, 1885   Anchor   329,255

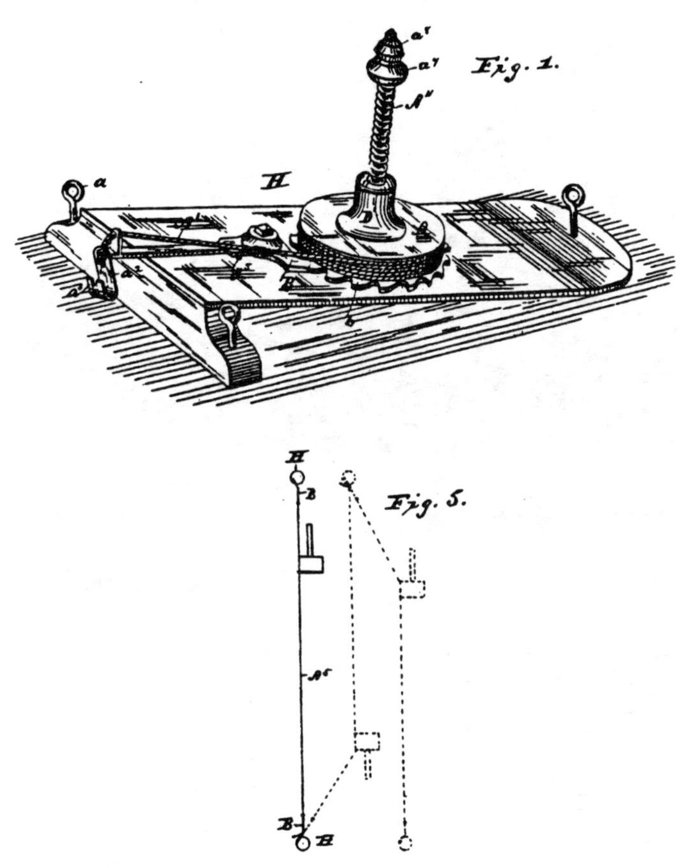

Quincy, Adams County, Illinois

This invention provides for tension control on the check-row cord or wire, and for automatic yielding to compensate for the lateral strain due to the position of the planter.

ROBERT FARIES

June 8, 1886   Anchor   343,305

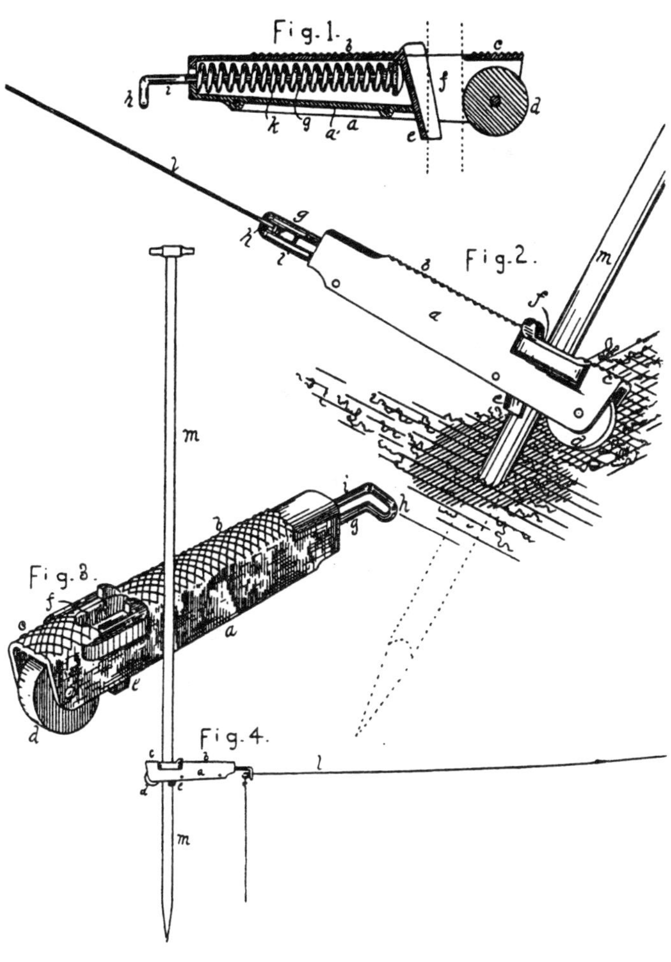

Decatur, Macon County, Illinois

ABIJAH T. CROW

August 24, 1886   Anchor   348,152

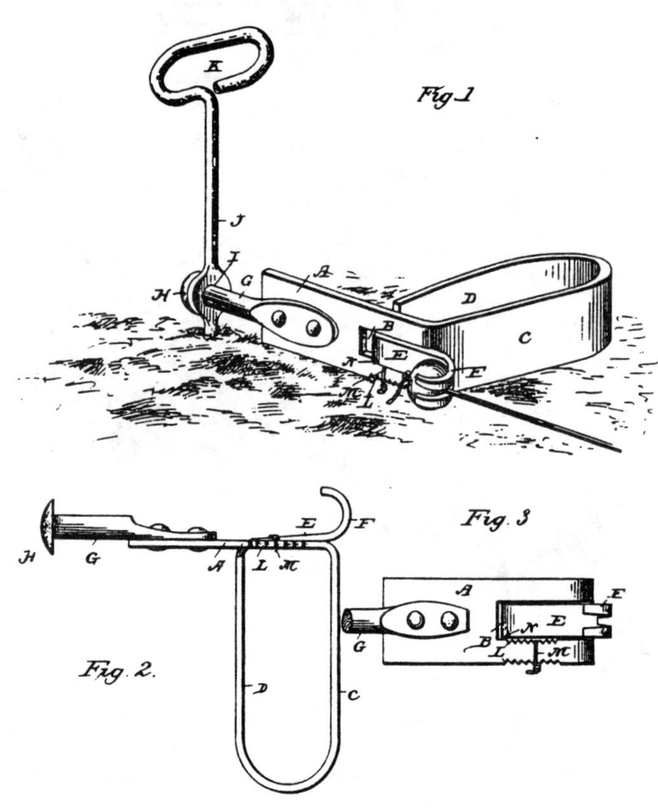

Clarinda, Page County, Iowa

February 22, 1887   Laydown/Takeup Machine   358,354

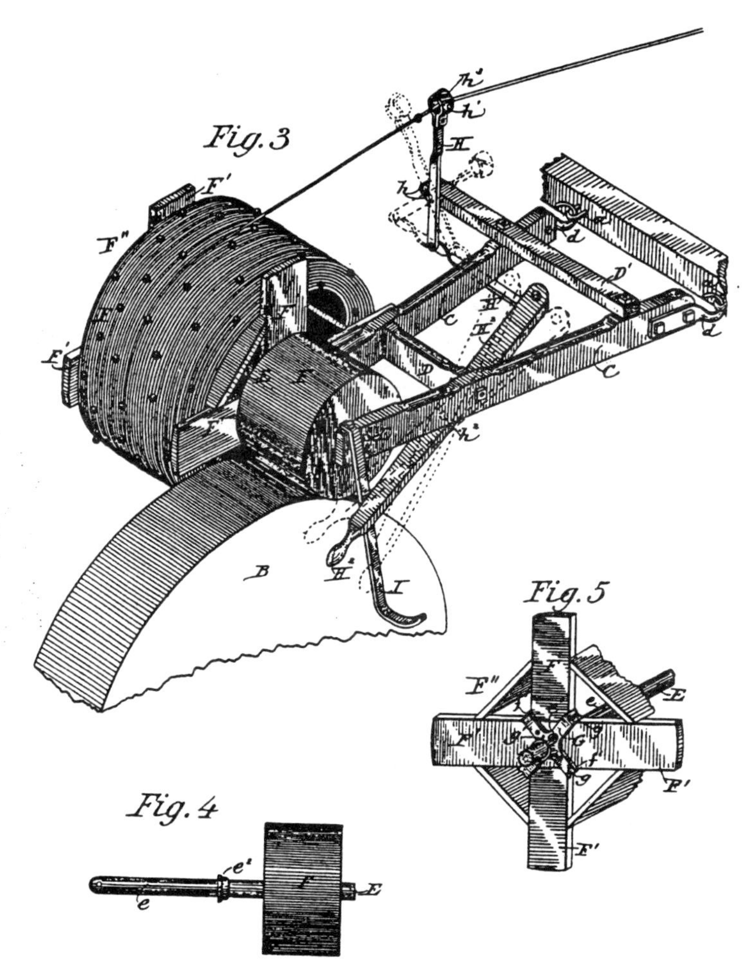

Galva, Henry County, Illinois

August 2, 1887   Anchor   367,582

Auto –Tension Extension

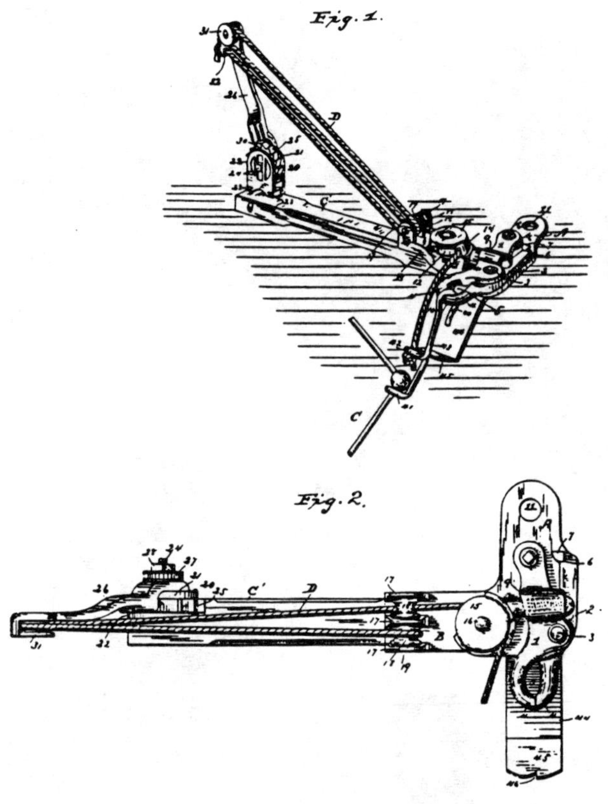

Quincy, Adams County, Illinois

JOSEPH C. BARLOW

November 15, 1887   Anchor   373,170

Improvement on 367,582

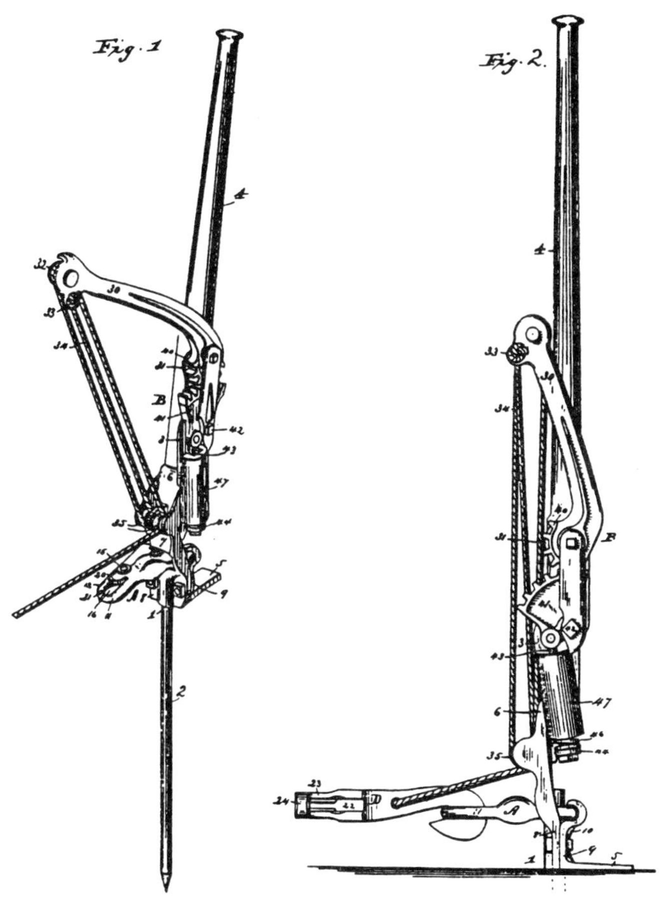

Quincy, Adams County, Illinois

February 26, 1889   Anchor   398,485

Moline, Rock Island County, Illinois

March 5, 1889   Self-Winding Reel   398,872

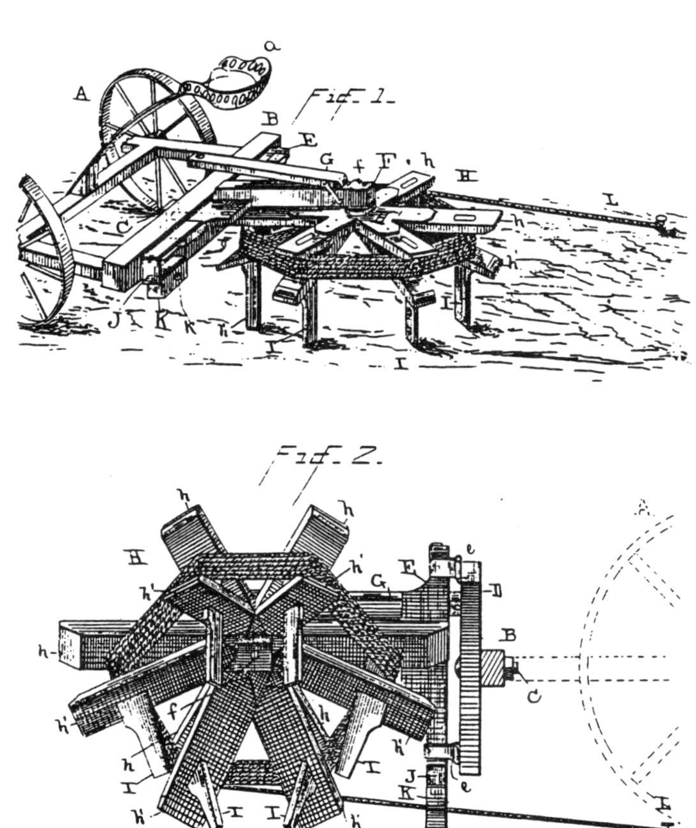

Red Oak, Montgomery County, Iowa

May 21, 1889   Marking-Chain   403,516

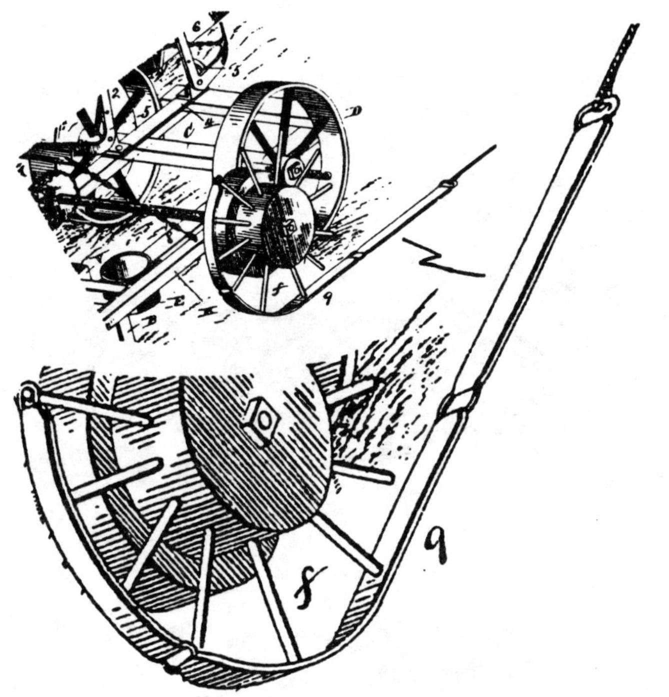

Blue Mound, Linn County, Kansas

"When it is desired to use the machine, short marking-chains *9* are secured at one side of the field and wrapped around the wheels as shown in Fig. 1. As the machine is drawn over the field, the spike-wheels will be caused to rotate by the chains, and the line of the row thus indicated. . . . and after the chain is run out, the spike wheel is caused to rotate by contact with the ground."

This process starts the drop mechanism in the same place for each row, much as would a check-line.

November 19, 1889   Reel   415,464

Aurelia, Cherokee County, Iowa

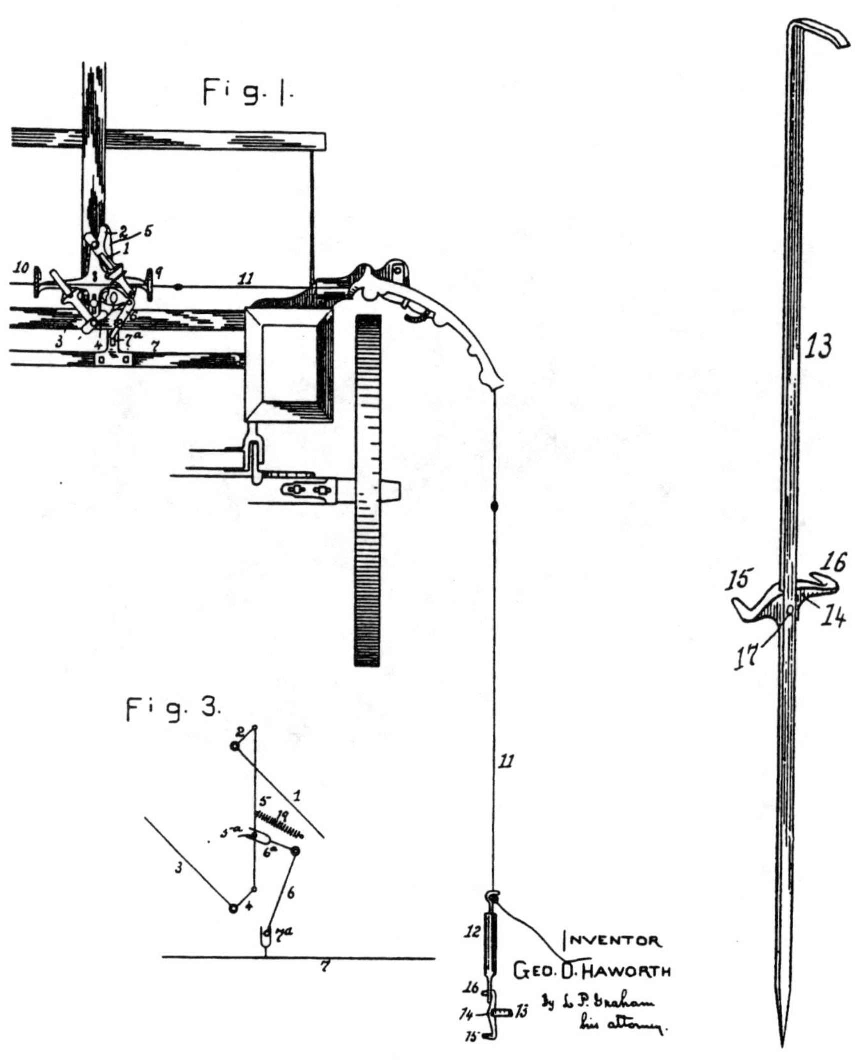

Decatur, Macon County, Illinois

JOSEPH C. BARLOW

July 7, 1891 · Anchor · 455,380

Improvement on 373,170

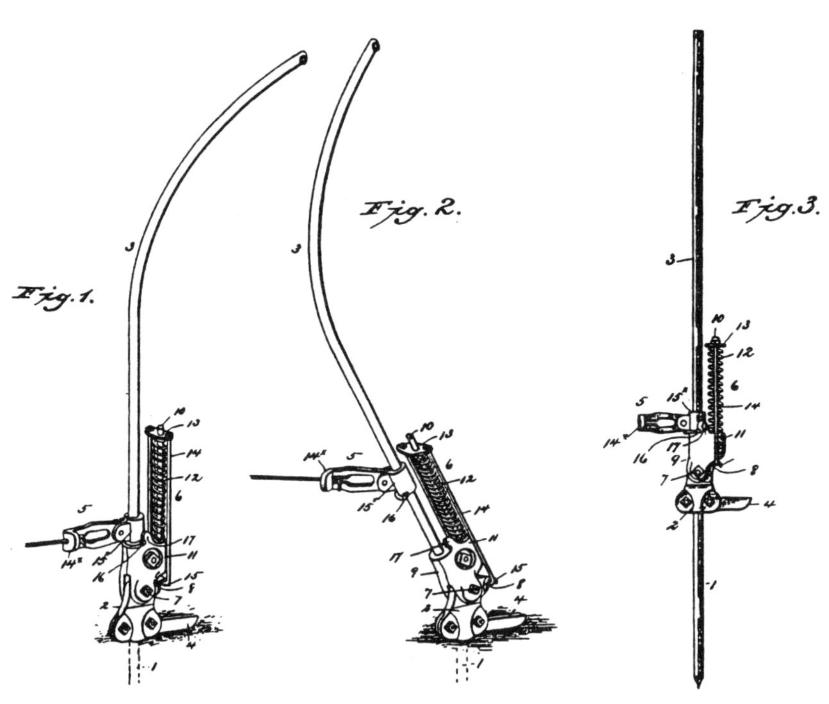

Quincy, Adams County, Illinois

December 5, 1893   Anchor   509,913

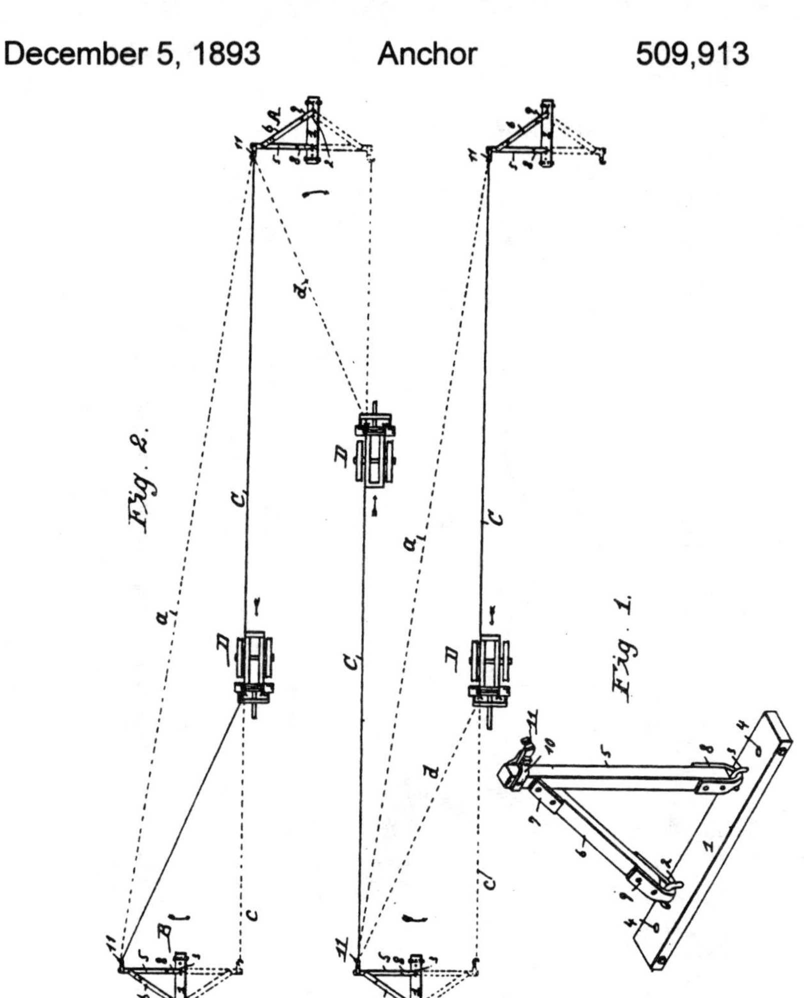

Lebo, Coffey County, Kansas

"This invention is designed for use as an anchor for check-row wires; and relates to that class where a swinging frame or arm is employed to automatically carry the wire to a position in line with the approaching planter, ..."

DANIEL AUMILLER

December 12, 1893   Anchor   510,463

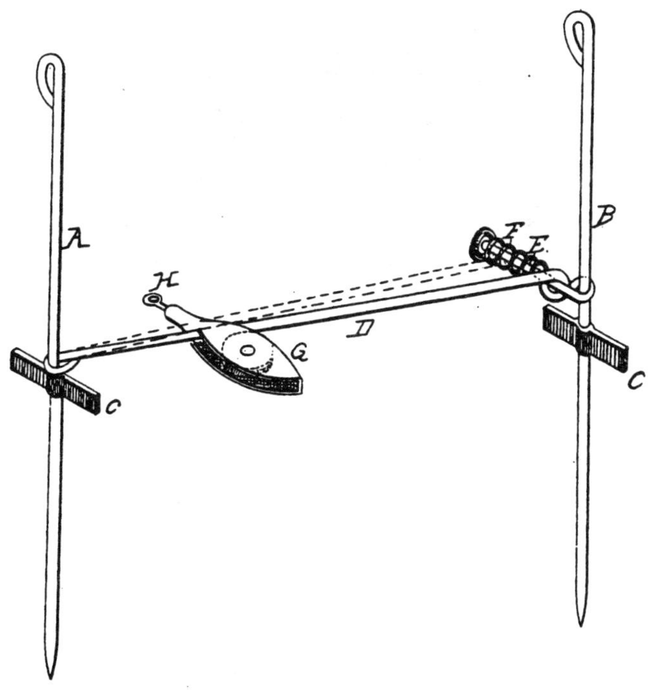

Bucyrus, Crawford County, Ohio

July 24, 1894   Anchor   523,388

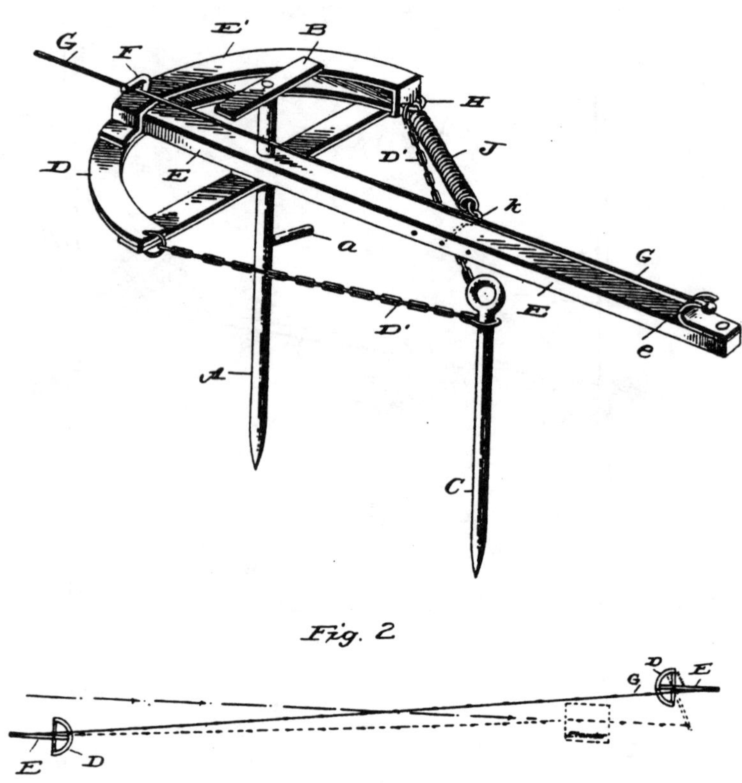

Aplington, Butler County, Iowa

September 25, 1894   Anchor   526,424

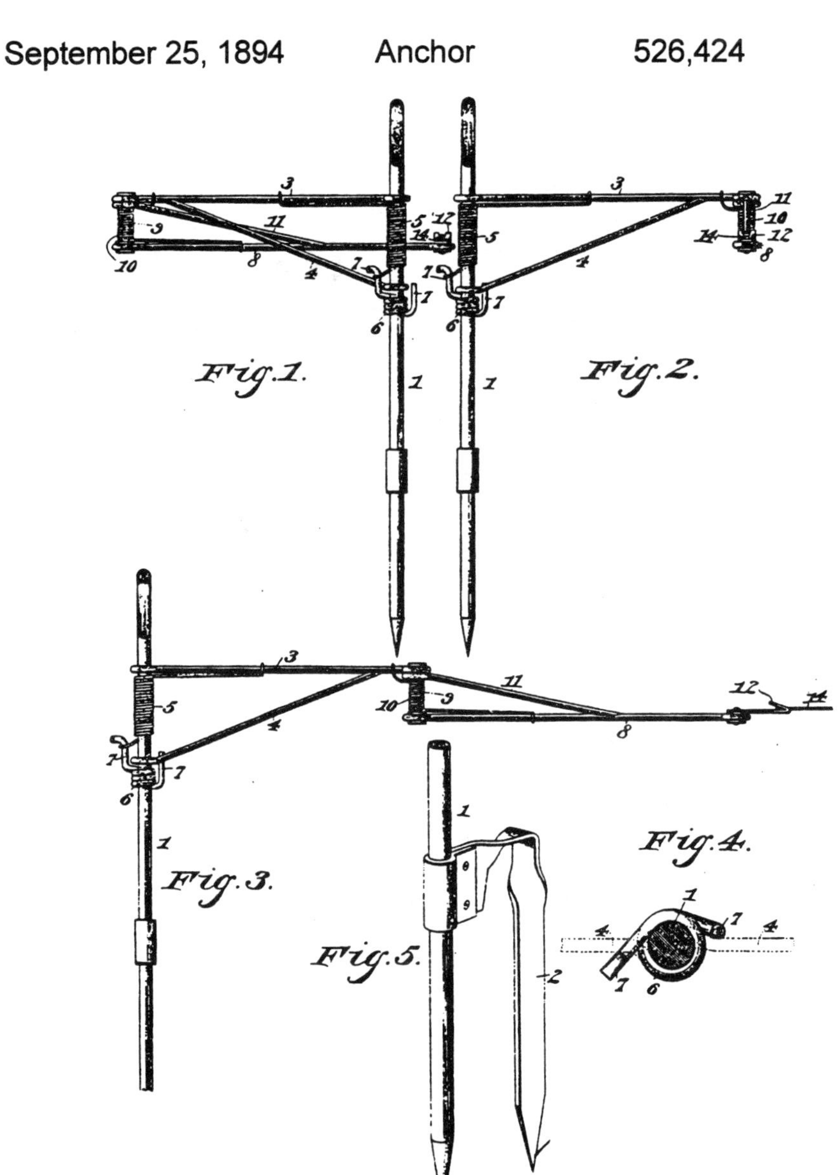

New Franklin, Howard County, Missouri

February 19, 1895   Reel   534,474

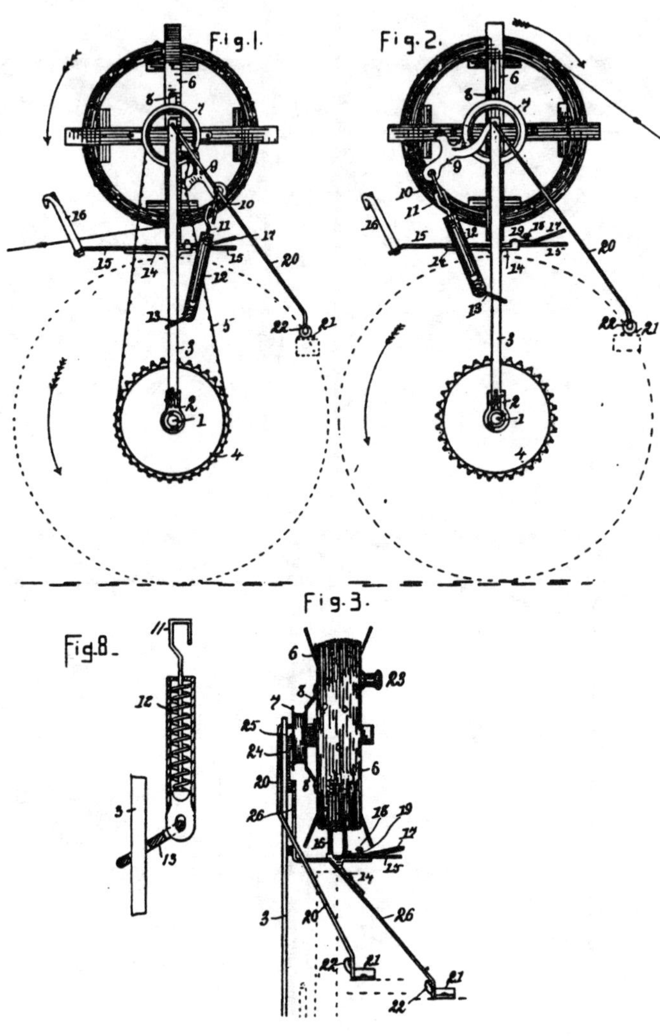

Decatur, Macon County, Illinois

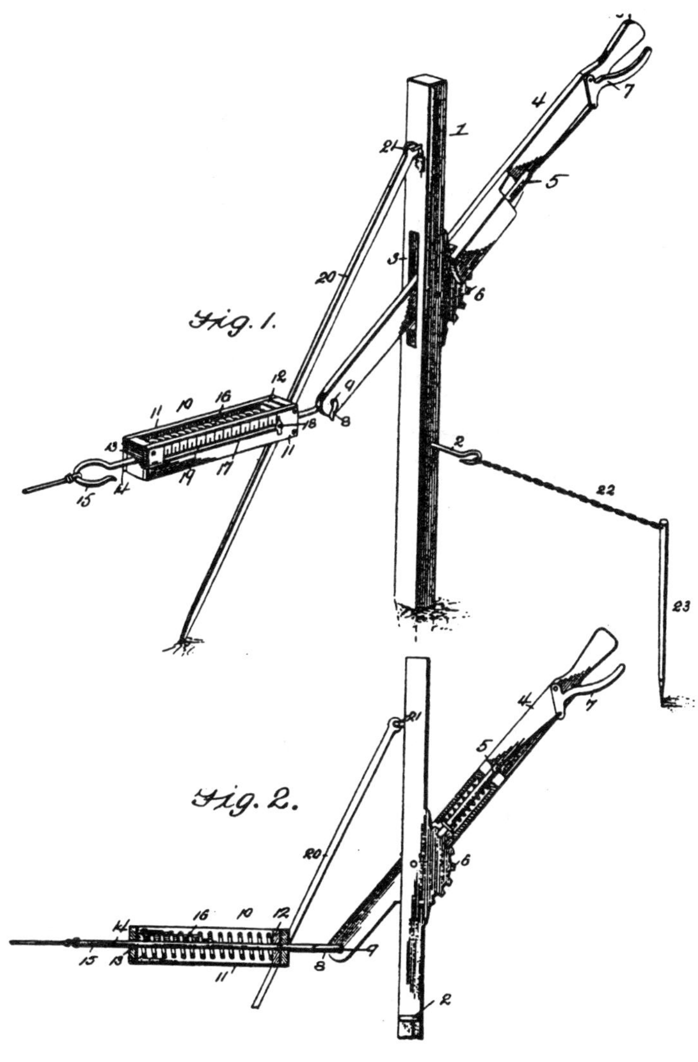

Sibley, Ford County, Illinois

MANNIN GROSS

July 2, 1895   Wire Stretcher   541,850

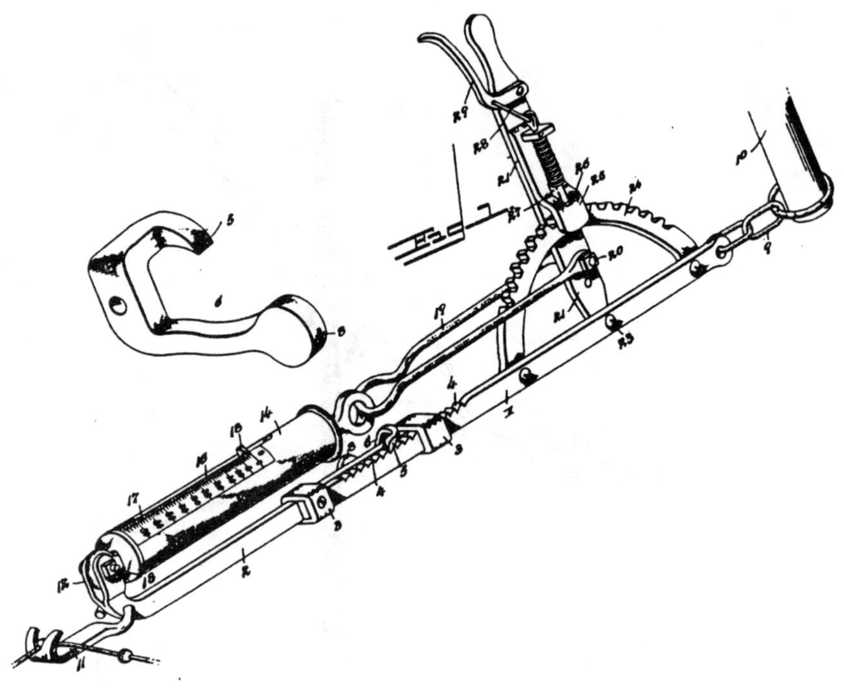

Burlington, Coffey County, Kansas

"This invention relates to wire stretchers; and has for its object to provide a new and useful device of this character that is especially adapted for use in stretching the check-row wires of corn planting machines to provide means for maintaining the check-row wire at uniform tension, whereby the planting may be done in accurate check-row."

"It is obvious that the herein described wire stretcher may be adapted for a variety of other purposes, such as for use in building wire fences, ..."

WILLIAM R. WAGGONER

January 26, 1897   Anchor   576,010

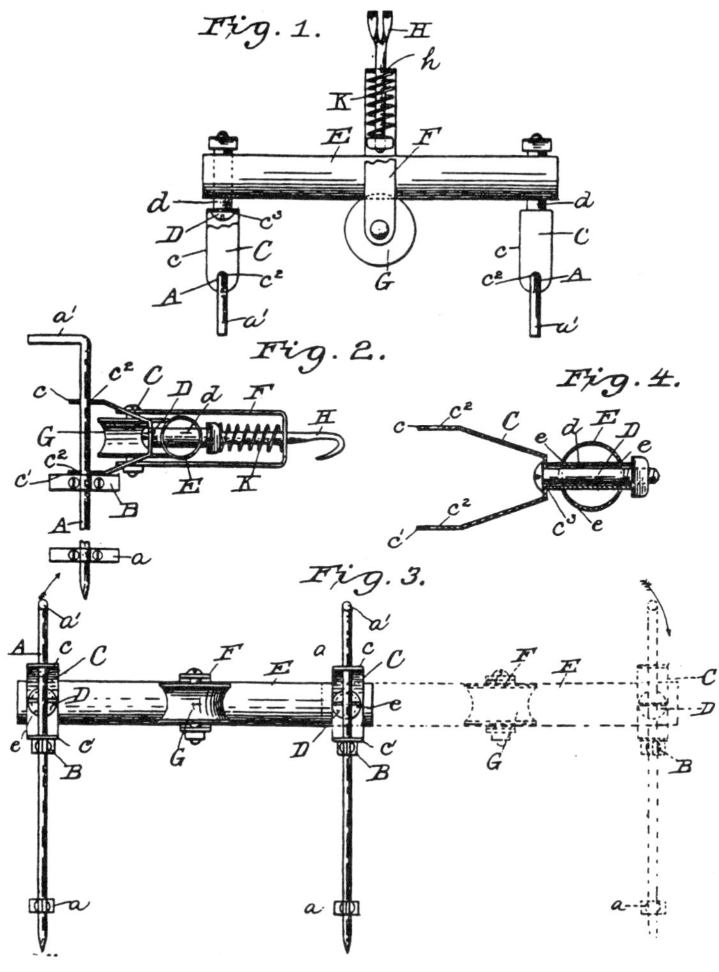

Art, Clay County, Indiana

February 14, 1899   Anchor   619,585

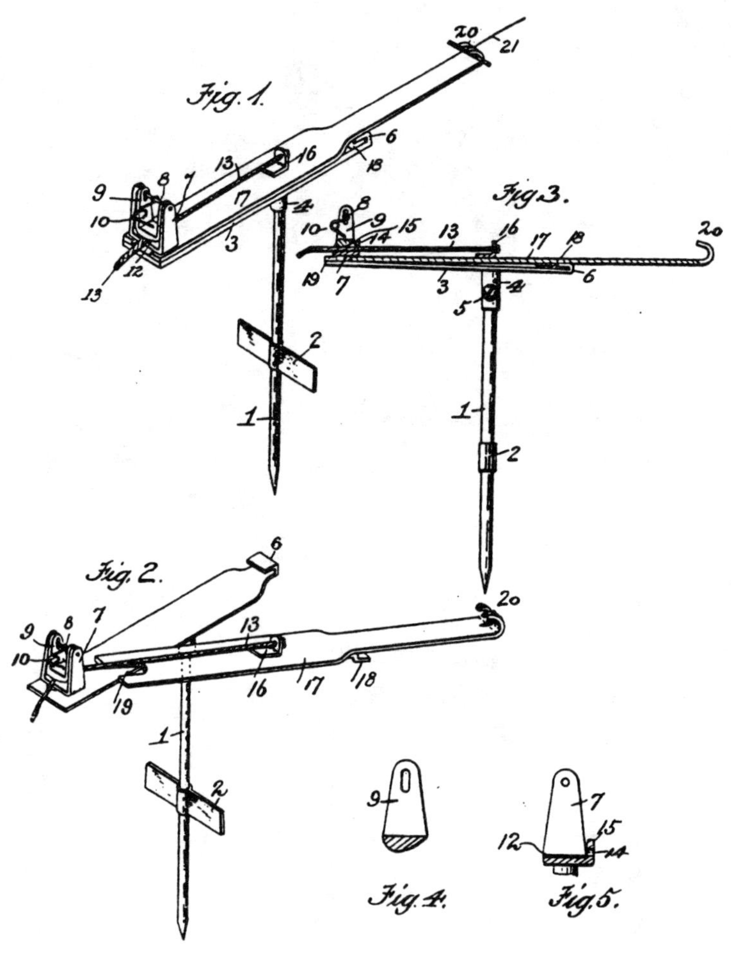

Newburg, Fillmore County, Minnesota

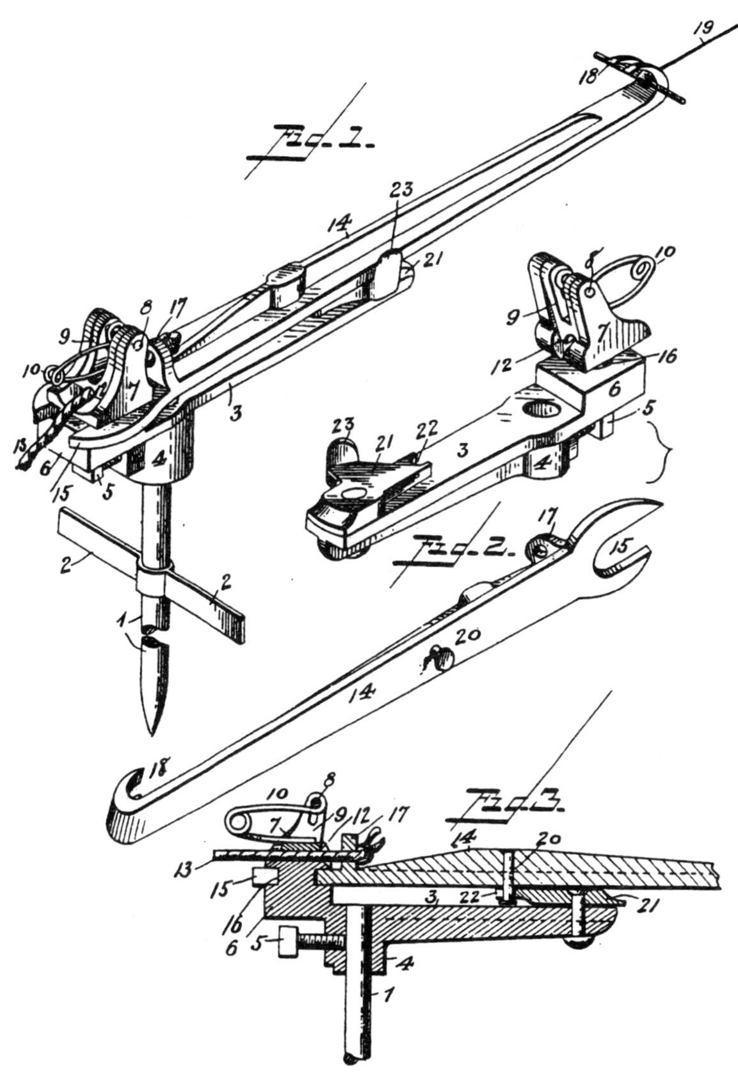

Newburg, Fillmore County, Minnesota

HERMAN W. WARREN

December 13, 1904   Chain   777,458

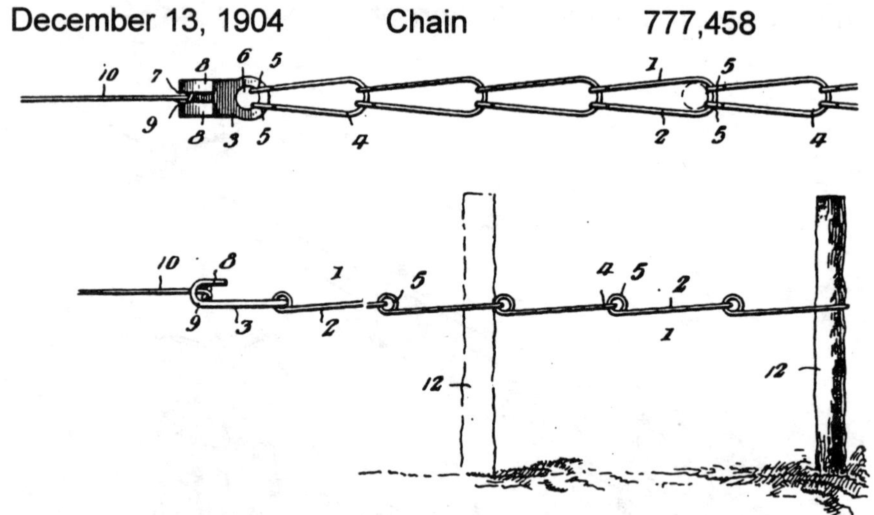

Sheridan, LaSalle County, Illinois

"When planting corn in a fenced field of irregular shape by means of a check-row planter, the check-row wire cannot always be anchored in the field being planted, as the securing means are connected to one of the buttons on the wire. The wire must therefore be passed through the fence before it can be anchored. This is often a troublesome and inconvenient operation and much disliked by farmers. My invention is to overcome this problem, and I claim:

A check-row wire attachment comprising a chain formed of elongated links, each link having an opening at one end for an anchor stake, and its opposite end connected to the following loop, and a hook one end of said chain having prongs adapted to engage a button on a check-row wire."

The chain is intended to be passed through the fence to meet the check-row wire, therefore not disturbing the wire's last position.

WALTER J. LUNDGREN

January 12, 1909   Tension Anchor   909,196

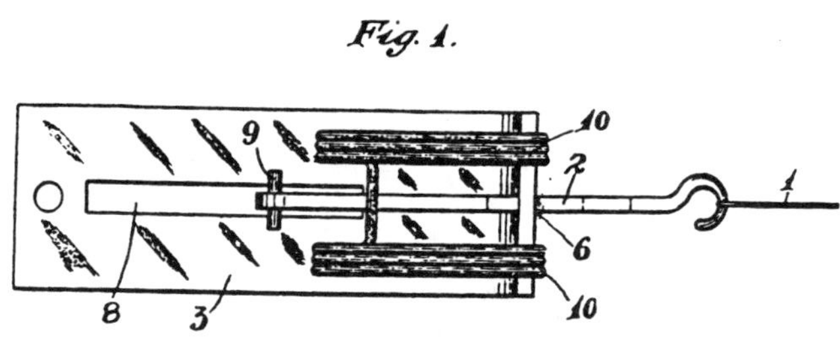

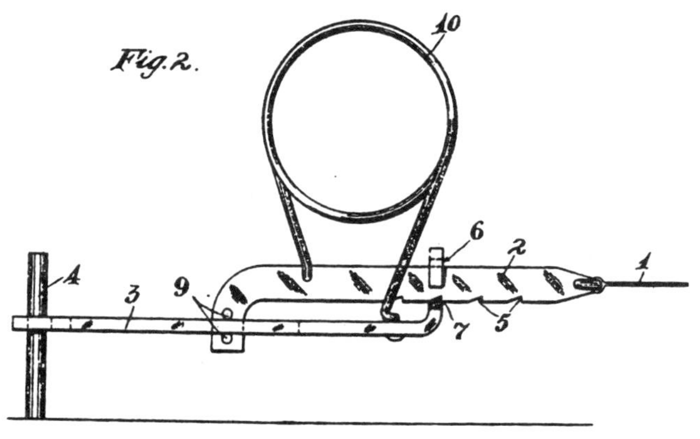

Willmar, Kandiyohi County, Minnesota

November 28, 1911   Tension Anchor   1,010,268

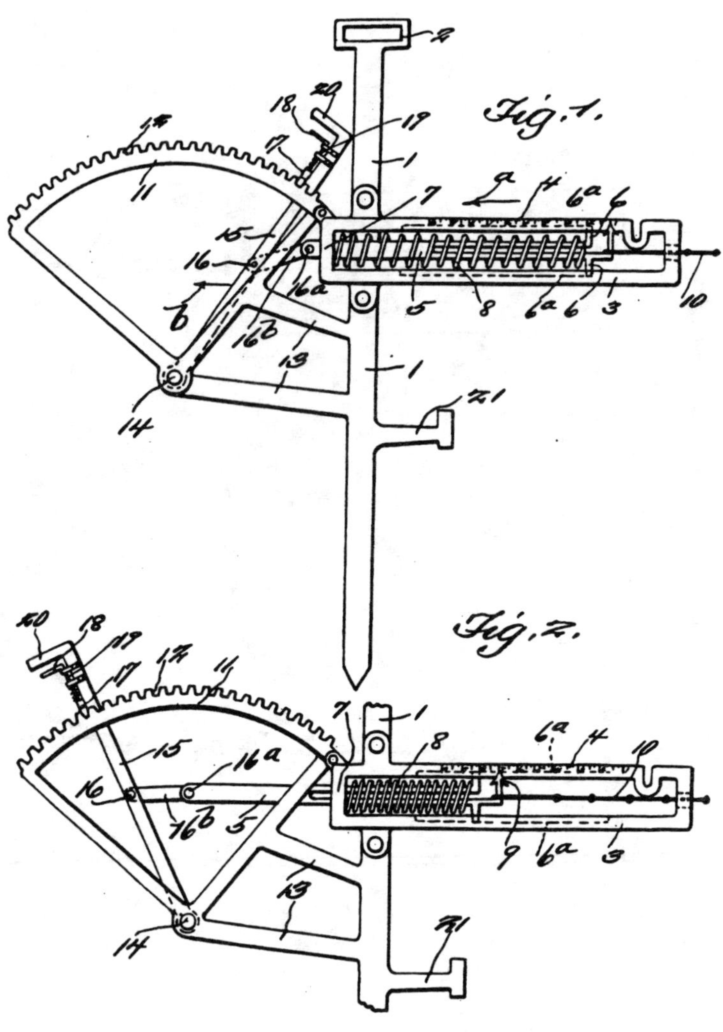

Greeley, Weld County, Colorado

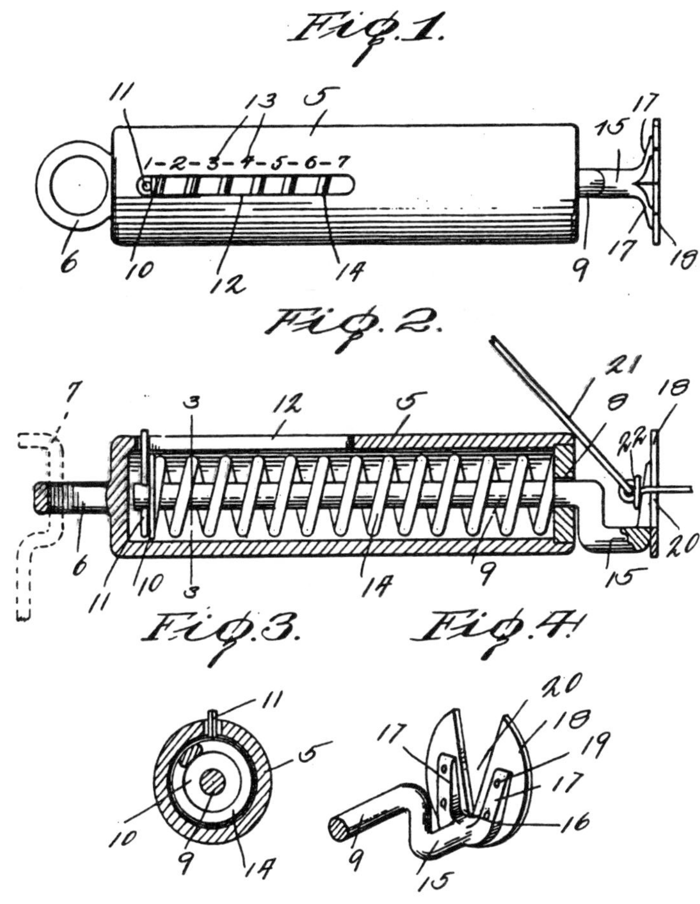

Princeton, Bureau County, Illinois

November 20, 1917   Tension Anchor   1,247,249

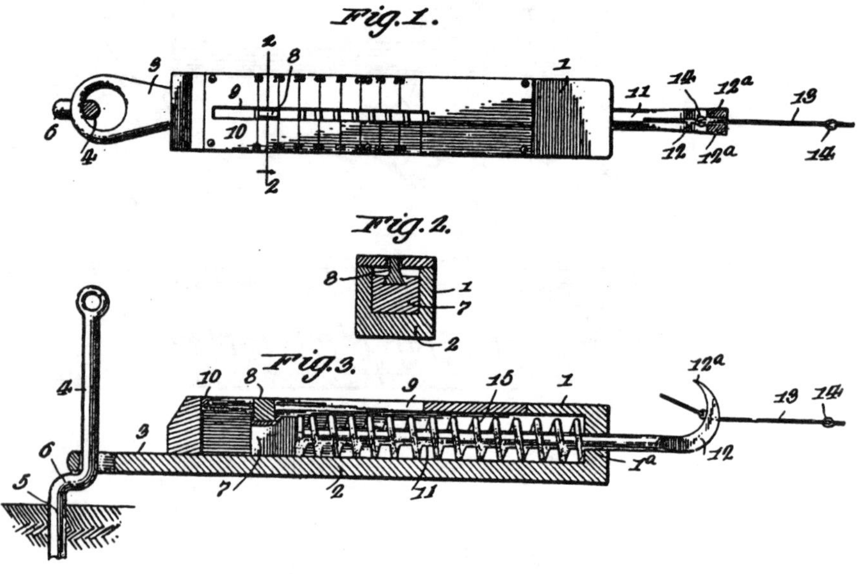

Ivanhoe, Lincoln County, Minnesota

March 5, 1918   Tension Anchor   1,258,102

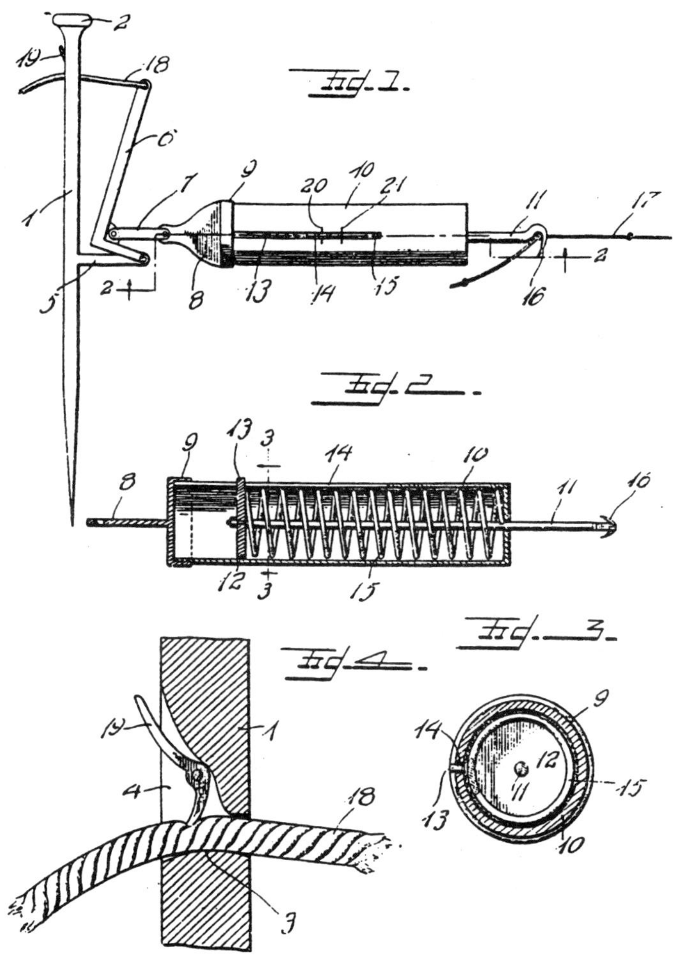

Augusta, Hancock County, Illinois

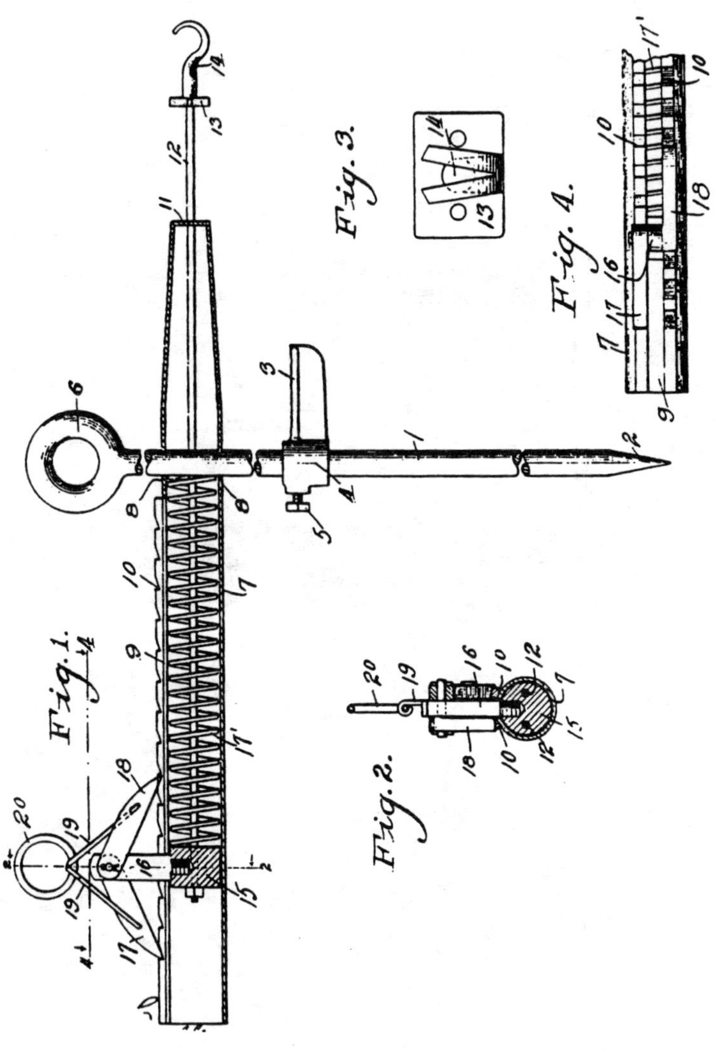

Fort Dodge, Webster County, Iowa

June 10, 1919   Tension Anchor   1,306,045

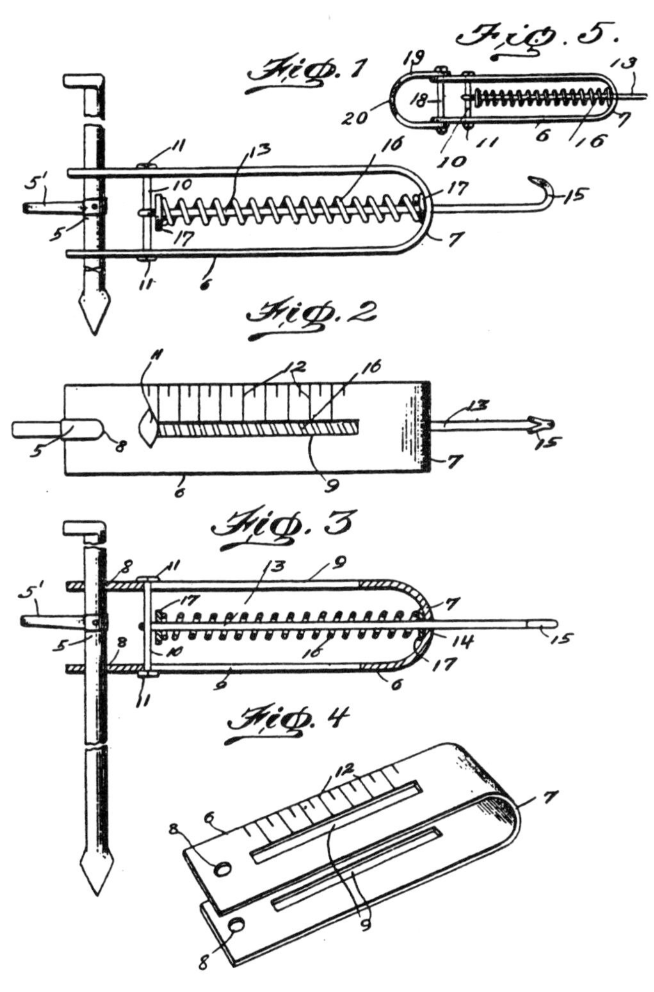

Wisner, Cuming County, Nebraska

June 19, 1923   Support Anchor   1,459,163

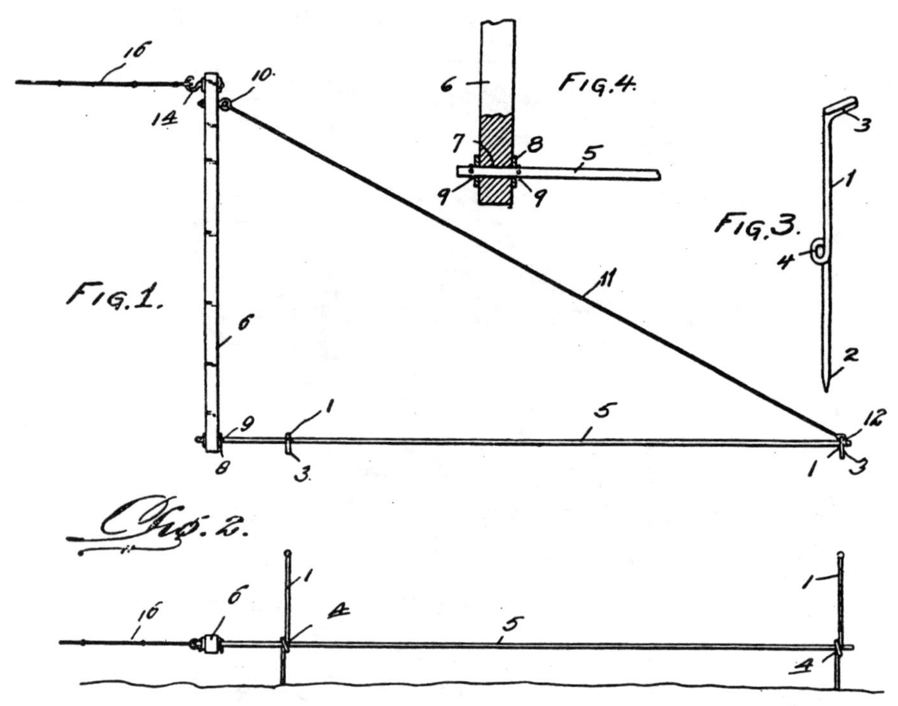

Moundville, Vernon County, Missouri

WILLIAM H. BORGERS

January 18, 1927   Tension Anchor   1,614,453

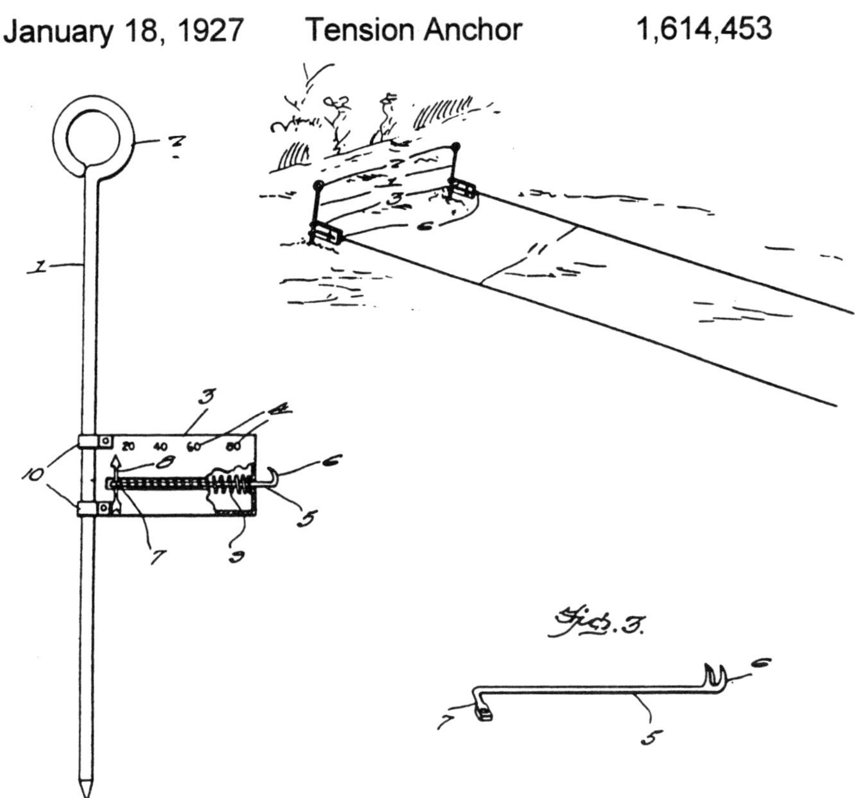

Marion, South Dakota

"This invention relates to an improved corn planter wire, and has for its principle object to provide an extremely simple and inexpensive gauge which serves to accurately indicate the degree of pull exerted upon the wire, whereby to permit the latter to be stretched sufficiently taut to provide a perfect guide in planting the rows of corn, thus eliminating the liability of crooked rows crosswise of the field."

The patent title is misleading calling it a "corn planter wire", when in fact it is a tension-gauge anchor for the wire.

SAMUEL K. DENNIS

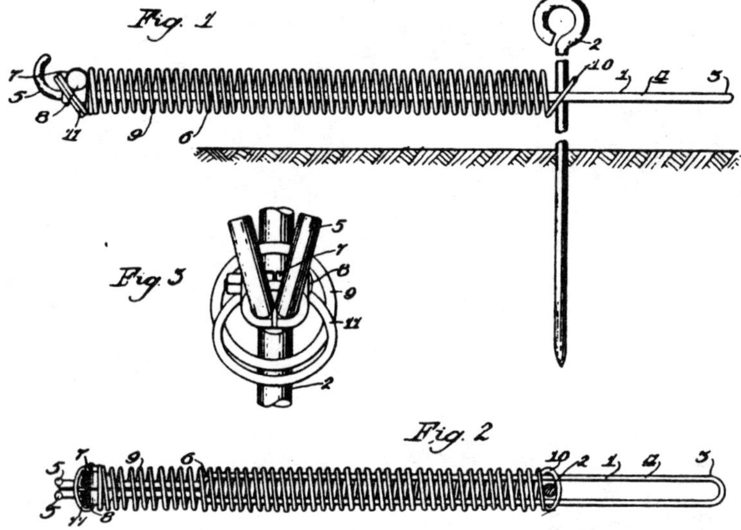

Oak Park, Illinois

Assigned to International Harvester

November 28, 1939   Wrapping Reel   2,181,441

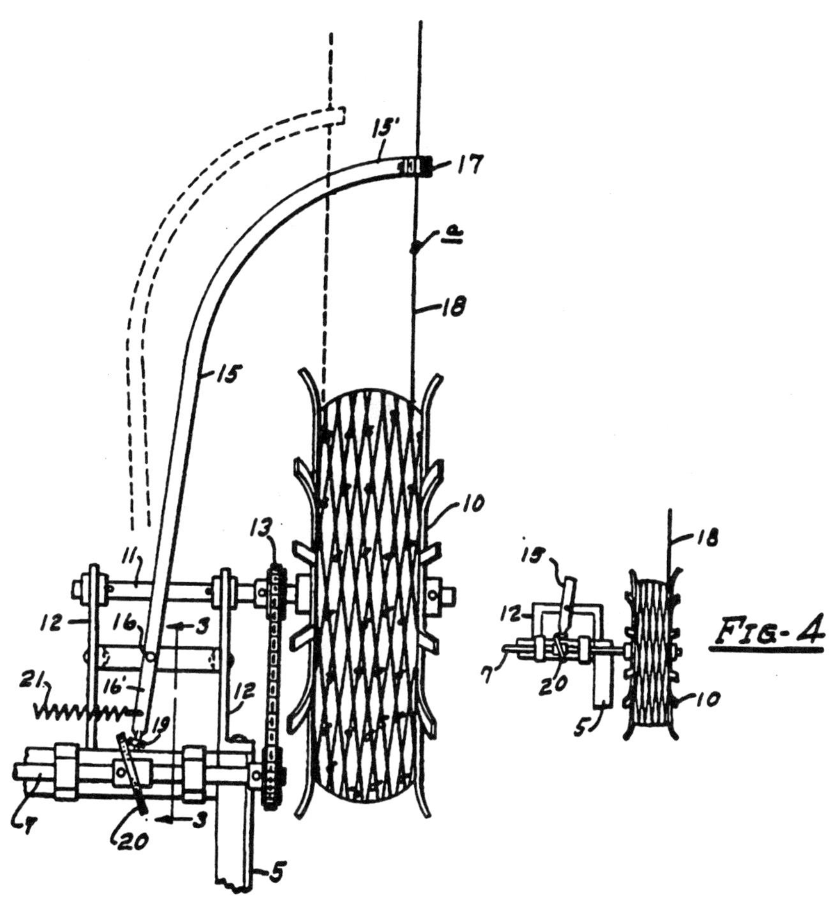

Rockford, Illinois

Assigned to J.I. Case Company

UNKNOWN FIND BY LARRY GREER

Anchor

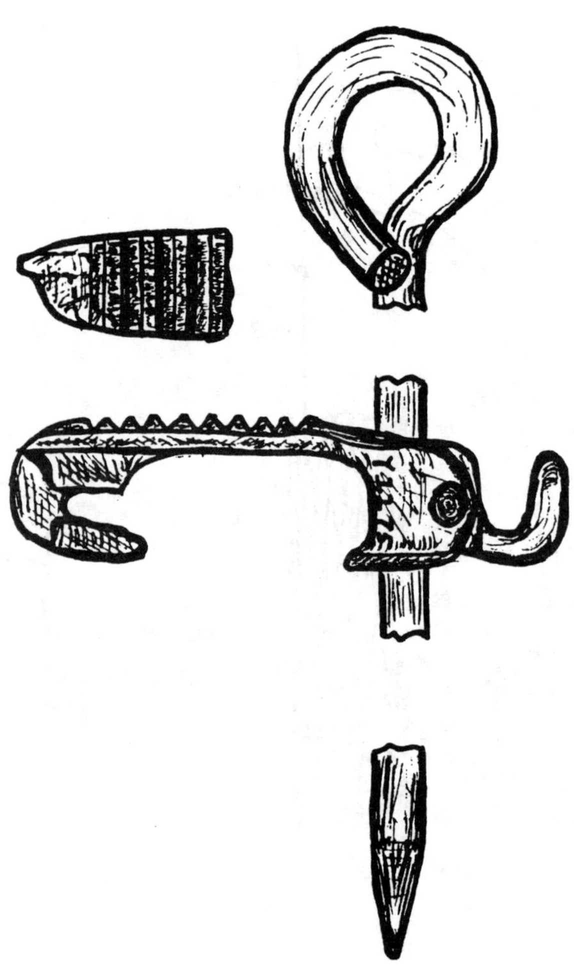

Lyon County, Kansas
March 18, 1974

Anchor

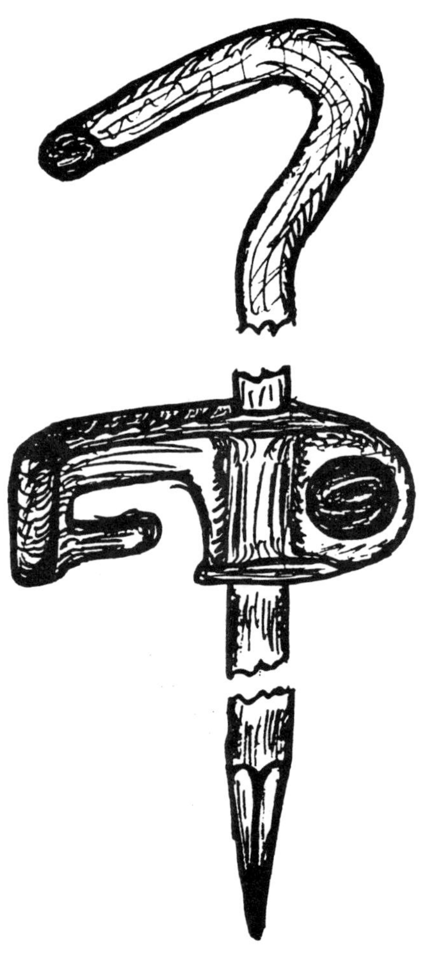

Lyon County, Kansas
June 27, 1974

SECTION 6

CHECK-ROW
PLANTER ADVERTISEMENTS
AND
OPERATION MANUALS

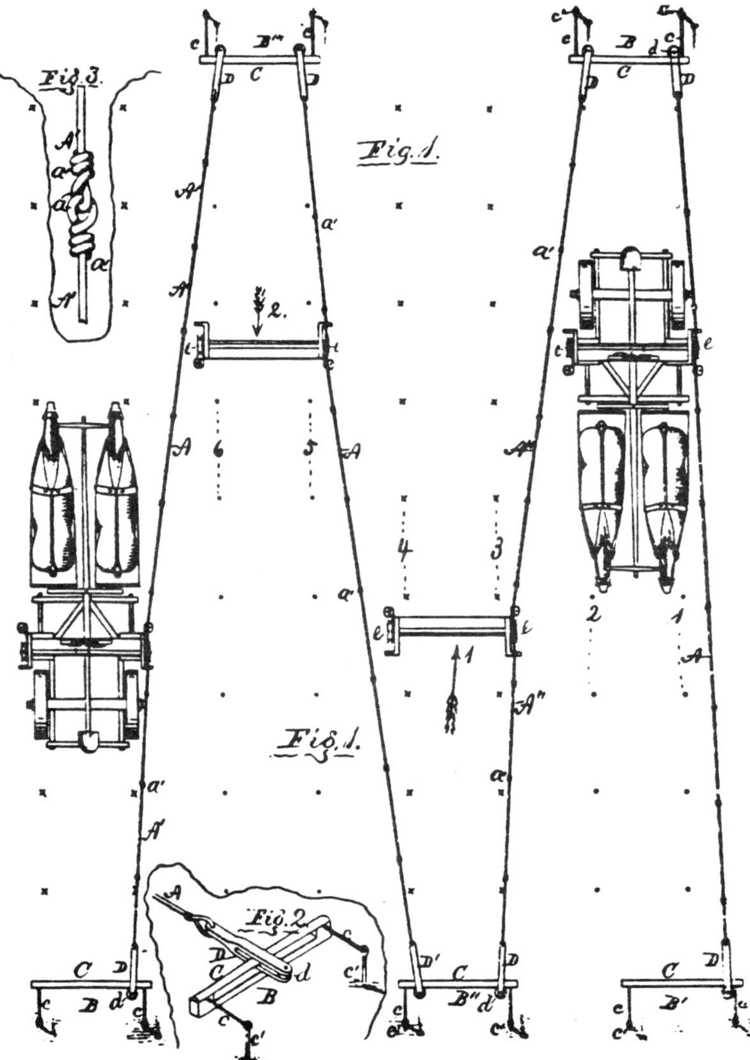

TAIT'S WIRE CHECK ROWER.
IS THE SIMPLEST AND EASIEST TO OPERATE THAT IS MADE.
I'M GOING TO HAVE ONE OF TAITS WIRE CHECK ROWERS
IF OLD JACK GETS THERE BEFORE THEY ARE ALL SOLD. GET UP JACK!
YOU BET JACK WILL GET THERE MA!
"I'm going to have one of 'Tait's Wire Check-Rowers,' if 'Old Jack' gets there before they are all sold. Get up, Jack!"
The above resolution was made by Farmer Johnson, after seeing how easy it was for his neighbor, Simpson, to plant his
corn straight with a "Tait Wire Check-Rower." So simple and easy to operate.
Manufactured by CONKLIN, TAIT & CO., Decatur, Ill.
TAIT'S WIRE CHECK ROWER.
HAS FEWER WEARING POINTS THAN ANY OTHER ROWER MADE
BY GOSH! PA HAS GOT IT.
Farmer Johnson returns with a "Tait Wire Check-Rower," and, you see, rejoices with his wife and young
"hopeful." No more crooked corn for him.
CONKLIN, TAIT & CO., Sole Manufacturers,
Decatur, Ill.
Grangeville, Iowa
FOUR YEARS OF CONTINUED USE OF THE TAIT ROWER HAS MADE FARMER JOHNSON
PROSPEROUS AND BROUGHT EASE AND COMFORT TO OLD JACK

| August 2, 1881 | Check-Row Planter | 245,028 |
| March 28, 1882 | C-R Attachment | 255,472 |

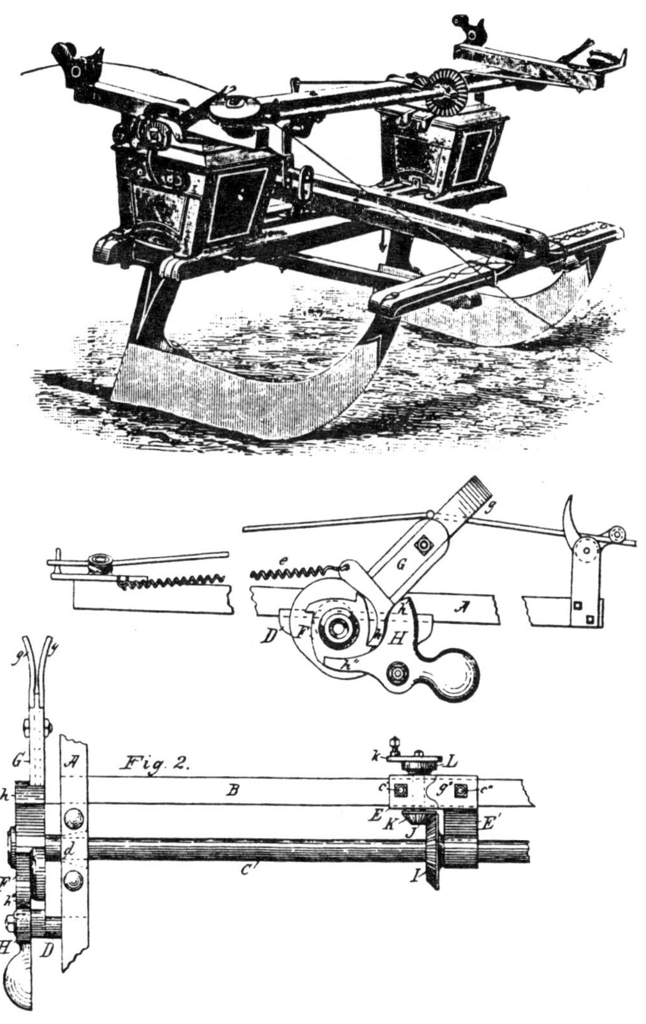

Decatur, Macon County, Illinois

AND GOOD WORK, THE
TAIT WIRE CHECK ROWER
IS AHEAD OF ALL OTHERS.

The unparalleled record made by the TAIT WIRE CHECK ROWER in the past, guarantees us in saying to the Implement Dealers and Farmers, that this Rower is the simplest and most durable, and easier to operate in the field than any rower in the market.

By a glance at the cut on the second page of this folder, you will readily understand its construction. A simple cog-wheel motion in the center, with an adjustable crank, gives it any length stroke — from one to four inches, all of which variations can be obtained by simply unloosening one set screw. These cog-wheels continue to revolve, and do not oscillate backward and forward like all other rowers, thereby insuring a uniform wearing of all bearings, and enabling the operator to take up any and all lost motion by merely loosening one set screw and lengthening the crank.

See its construction, and you will say it is the simplest and most powerful rower you ever saw.

It is guaranteed to check corn as well as can be done by hand, and as good as any rower.

Ask your dealer to see the TAIT Wire Check Rower.

CONKLIN, TAIT & CO.,

Sole Manufacturers, Decatur, Ill.

Decatur, Macon County, Illinois

"No more money paid out for Check Rower Extras.'

"IT'S GOOD ENOUGH FOR ME."

'" The Best and Simplest Rower made."

The above declarations are made by Farmers
who use the

Tait Wire Check Rower.

To see it is to pronounce it good.

To work with it is to make yourself happy.

Every knot is a coupling, which enables you to uncouple the wire at any joint and use only sufficient wire to plant your field, and thereby not be bothered with dragging extra wire after you.

We furnish pinchers for opening and closing these couplings.

We use nothing but the best annealed steel wire we can get.

Don't buy until you have seen the **Tait Wire Check Rower.** Some of the dealers in your town have it.

ASK FOR IT. If they do not have it, write

CONKLIN, TAIT & CO.,

DECATUR, ILL.,

Sole Manufacturers of Tait's Wire Check Rower.

Decatur, Macon County, Illinois

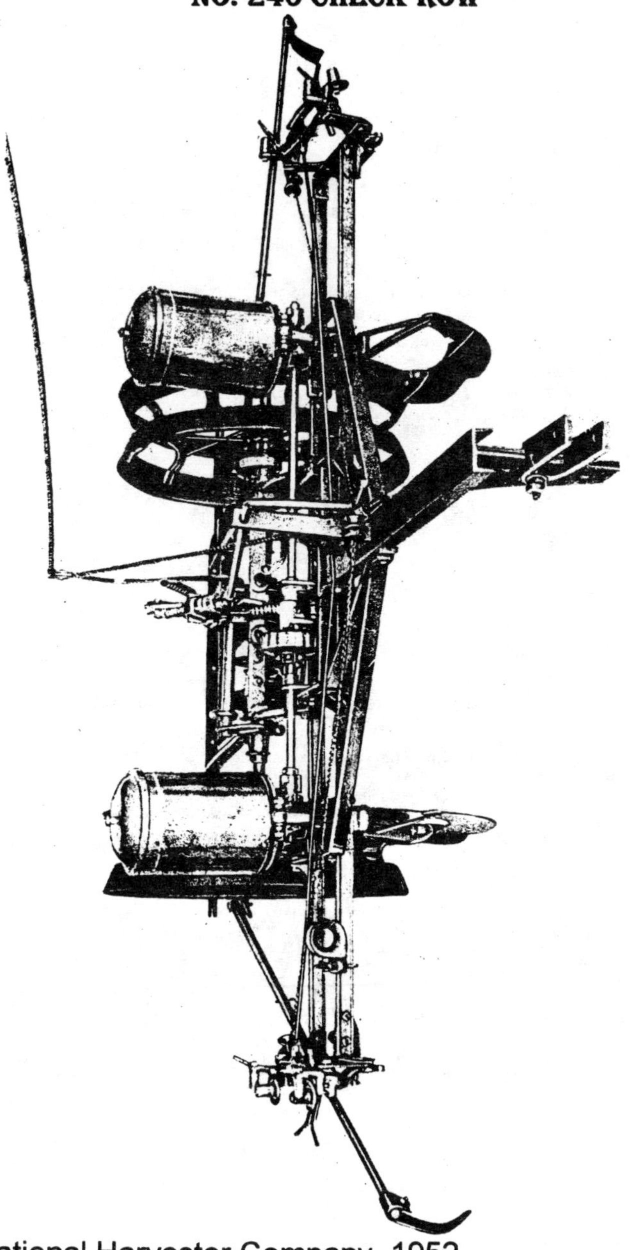

International Harvester Company, 1952

McCORMICK TWO-ROW CORN PLANTER
NO. 240 CHECK-ROW

LAYING OUT WIRE

The reel can be used on either side of the planter. In laying out the wire, drive the planter to the edge of the field and place it in position to drive across the field where the first two rows are to be planted. Take the wire from the underside and directly off the reel. Hook the wire on the anchor stake, leaving a few extra links back of the hook, and set the anchor stake to the rear of the planter about twice the distance between the rows and out of line with the center of the planter toward the near edge of the field. Apply tension on the spring holding the friction wheel in contact with the planter wheel.

Drive carefully and straight across the field so the wheel marks can be used as a guide on the return trip when planting the first two rows.

At the far end of the field detach the reel and turn the planter around and into position for planting the first two rows. Drive the planter far enough into the field to allow about five links of wire between the back of the planter and the point where the stake will be set. Remove enough wire from the reel to allow about four extra links of wire behind the stake, hook the wire to the anchor stake, draw it fairly tight, and set the stake in line with the wheel of the planter. In setting the stake, see that the wire has a uniform tension after the stake is in the ground.

NOTE: The stakes must be set the same distance behind the planter at each end of field.

STARTING TO PLANT

Be sure everything is in proper adjustment and in good running order before starting to plant. Set the marker on the side toward the open field, taking care that it is properly adjusted to conform with the width of the rows. *See Illusts. 3 and 4.* Place the wire in the check arm fork, making sure the fork prongs have the proper adjustment (they should be parallel and have about a 3/16" opening between them); thread the wire into the check head and close the check head. Lower the openers into the ground and adjust to the depth desired by means of the adjusting crank.

In making the first trip across the field with the planter in operation, follow the wheel marks made while laying out the wire. Always drive at a steady speed. Do not drive the planter nearer to the end of the field than five links of the wire. This will leave room for planting four head rows or one round across each end of the field and will avoid unnecessary strain on the wire. When planting the first time through the field make sure when turning at the end of the field that the wire releases from the check head before turning too far. After the first time through there will be enough angle in the wire to release it.

International Harvester Company, 1952

McCORMICK TWO-ROW CORN PLANTER
NO. 240 CHECK-ROW

Figure 1—Laying out the wire.

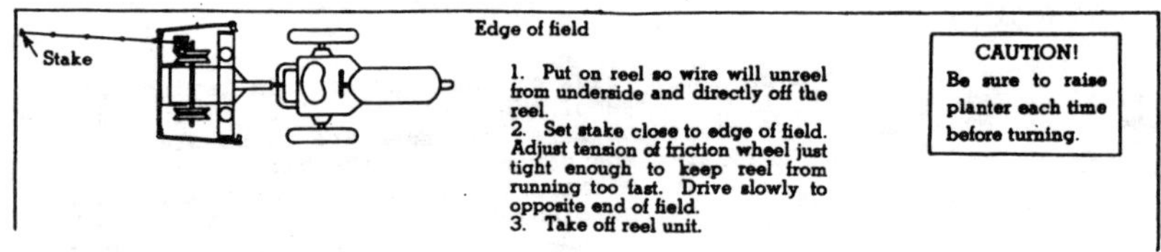

Figure 2—Starting to plant.

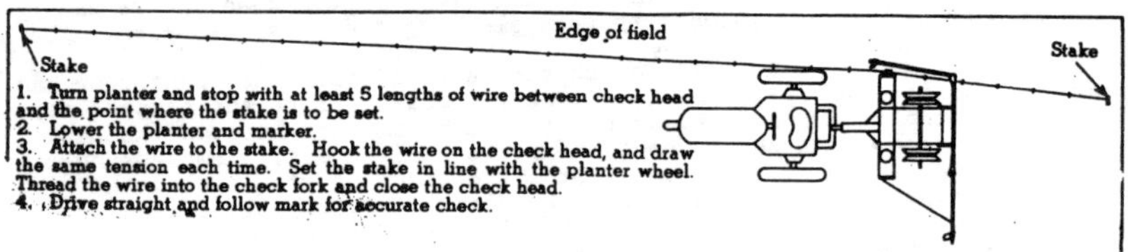

Figure 3—Checking.

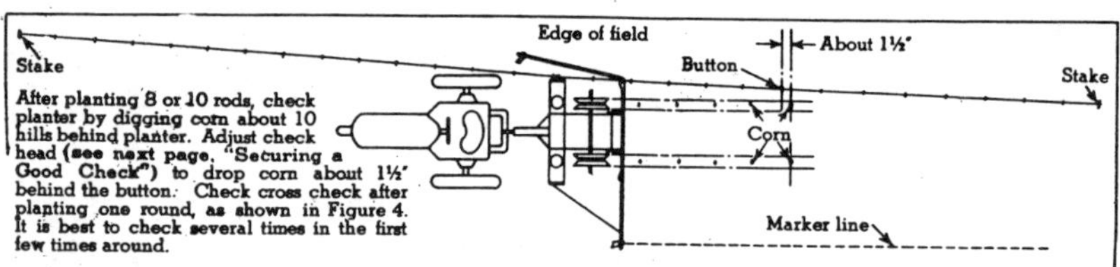

Figure 4—Turning at ends of field.

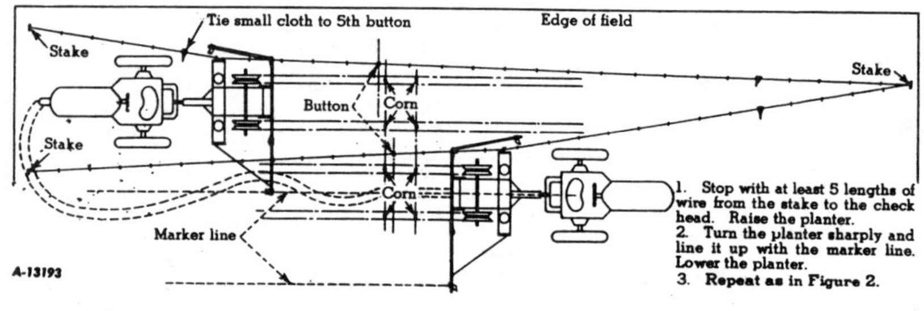

Illust. 21
Method of laying out the wire.

International Harvester Company, 1952

SECURING A GOOD CHECK

See Chart on page 22.

To be sure of a good check, it is advisable, after having planted a few rows, to dig up several hills across the rows out some distance from the end of the field and see whether or not they are in check.

When digging hills of corn to determine the accuracy of the cross check, always dig crosswise of the row and be sure that the corn is not moved. The hills should be located approximately 1-1/2 inches behind the button on the check wire. *See Illust. 21, Figure 3.*

Should the hills show out of check, the distance of this offset is twice the amount of the check head adjustment required to correct the offset. For example, if the hills are being carried too far in the direction of travel, move the check heads *(see Illust. 22)* forward one hole at "A" and lengthen the links to correspond at "B". If the hills are not carried far enough forward in the direction of travel move the check heads *(see Illust. 22)* rearward one hole at "A" and shorten the links to correspond at "B".

Be sure to adjust both the check head (A) and the link (B), when making check head adjustments.

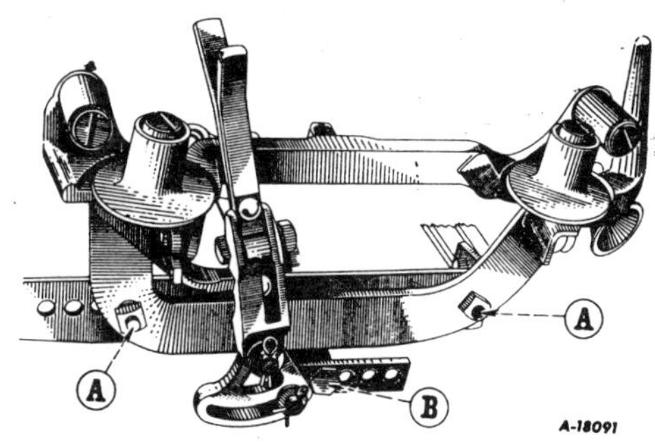

Illust. 22

Minor adjustments may be made to carry the hills forward by lowering the clevis to throw the heel of the boot slightly forward; or if the hills are carried too far, the clevis may be raised to throw the heel of the boot slightly rearward.

TAKING UP THE WIRE

Place the complete reel unit on the planter, thread the wire through the guides and fasten the wire to the reel so that it feeds over the top of the reel and sheave. Tighten the hand wheel on the reel tension bolt so that the friction wheel will keep the wire tight in front of the reel when driving towards the anchor stake across the field. *See Illust. 47.*

International Harvester Company, 1952

CHECK-ROW WIRE REEL UNIT

The reel unit can be assembled on either the left or the right side of the planter.

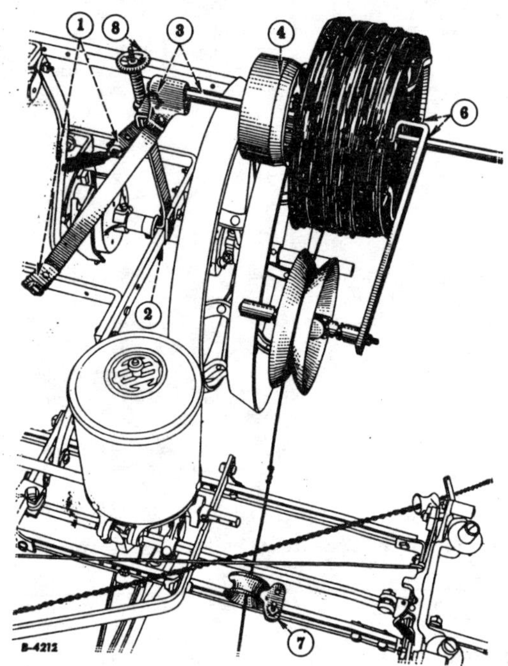

Illust. 47
Check-Row wire reel and axle assembled on the left side of the planter.

NOTE: Illustration shows the reel assembled in the correct manner for reeling up the wire.

International Harvester Company, 1952

KEEP THIS BOOK FOR FUTURE REFERENCE

INSTRUCTIONS

FOR

SETTING UP AND OPERATING; ALSO
REPAIR PARTS LIST FOR

CASE

No. 32-G PLANTER
(Check Row and Drill)

J. I. Case Company, Inc.
Established 1842
Racine, Wisconsin, U. S. A.

Form 879 1500-9-29-HPCo.

Printed in U.S.A.

Racine, Wisconsin, 1929

J.I. CASE COMPANY CHECK-ROW AND DRILL

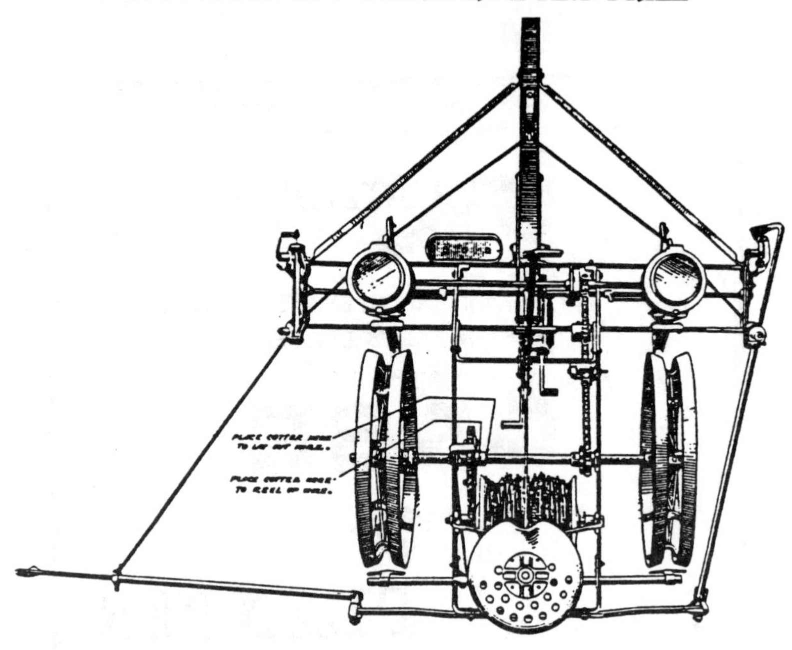

Fig. 8
(TOP VIEW)

INSTRUCTIONS FOR OPERATING

To Lay Out Wire

Place one anchor stake in corner of field and attach wire to stake and line up outside check head with stake.

Place reel in frame so that the wire runs off from the top side.

Remove cotter through reel drive sprocket assembly and main axle so sprocket assembly will revolve on main axle. (See Fig. 8).

Use one of the anchor stakes as a brake for the reel, by holding same in convenient place between frame and reel, so that reel will not pay out wire faster than the planter is travelling.

To Reel Up Wire

RUNNERS SHOULD BE OUT OF GROUND WHEN REELING UP WIRE.

Drive team straddle of wire or so that wire comes under pole. Place cotter in reel drive sprocket assembly and axle so that sprocket will rotate reel. Thread wire in guide hook on under side of front end of pole and through the wire guide in center of runner frame. Connect end of wire to reel so it will run on from under side.

Use one of the anchor stakes as a guide for wire when reeling up. Place lug on stake through hole in seat bar with loop on the under side ahead of axle.

IMPORTANT

In reeling up wire, it is necessary that the Adjusting Springs for reel sprocket friction plate be adjusted so that this may slip, in order to compensate for larger diameter of reel as wire is picked up. Adjust the 3 bolts for springs so that there is just friction enough on place to take up slack in wire on the ground.

"The Hamilton" Corn Planter.

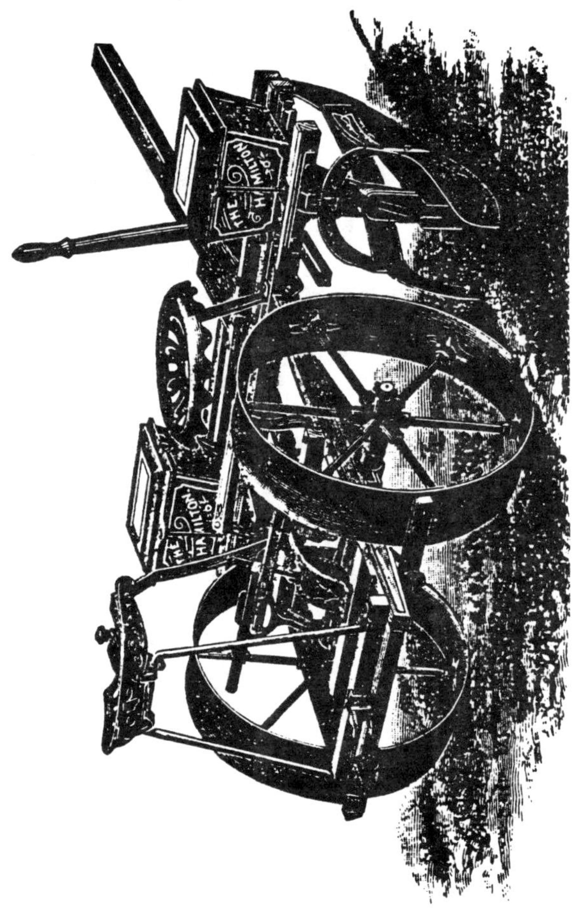

Showing the wheels set to run OFF of the seed row.

Hamilton, Ohio, 1891

THE "UNIVERSAL" CHECK-ROWER

Is the Least Complicated, and the Most Common-Sense Check-Rower on the Market.

The "Universal" is superior to all other check-rowers on account of its simplicity and freedom from delicate complicated machinery. It can be thrown in or out of gear instantly, without moving the wire, allowing the driver to stop the check-rower any time desired in the field.

Being of so few parts it cannot well get out of working order. The wire does not cross the Corn Planter, thus avoiding a great wear and strain on the wire and friction on the pulleys. A wire that does not cross the Planter will outwear several wires that do cross.

With the Check-rower the ground can be planted as fast as it is plowed, instead of having to wait till the field is all finished and marked out before planting can be commenced.

A farmer with the same amount of help can put in a much larger crop in the proper season.

It plants corn straighter than by any other method, and the advantage of having straight rows will be appreciated by all farmers, who know how much easier it is to plow out the weeds between straight rows than if the rows are crooked and irregular.

Hamilton, Ohio, 1891

THE PERFECT SPOOL
WIRE WINDER

A Product of

WIRE WINDER MFG. CO.
Mendota, Illinois

Mendota, Illinois, circa 1940

WIRE WINDER MFG. COMPANY

INSTRUCTIONS

1. Place mounting bracket on top of left front spindle, bolt clamp to steering arm. Attach carrier to lower end of mounting bracket.

2. To assemble Wire Winder insert bolts in holes indicated by arrows in above illustration. The assembled view shows Winder in position to roll up wire.

3. Barb or Corn Planter wire is either rolled up or unrolled by driving forward. Easily watched from the left side of your tractor.

4. For rolling up wire the drive spool part P2 is pressed against left front wheel with aid of spring tension. Apply reasonably strong tension which acts as automatic clutch. When rolling up old rotten wire lessen spring tension to prevent breakage.

5. To unroll wire replace drive spool part P2 with Pay Out Sleeve Part P5 which also acts as a brake. Do Not Attempt to use Drive Spool P2 for unrolling spool of wire. To unroll wire drive tractor forward and wire unrolls from bottom of spool.

6. To unroll New Factory Spool of wire use Pay Out Sleeve Part P5 and then Part P13 is placed between Part P5 and new spool of wire.

7. Use Carrier Arm when transporting wire by releasing spring tension and place shaft over end of Carrier Arm Part P6.

8. In abnormal conditions it is recommended to decrease air in right front tire to prevent slippage on drive spool. Grease moving parts before using.

Mendota, Illinois, circa 1940

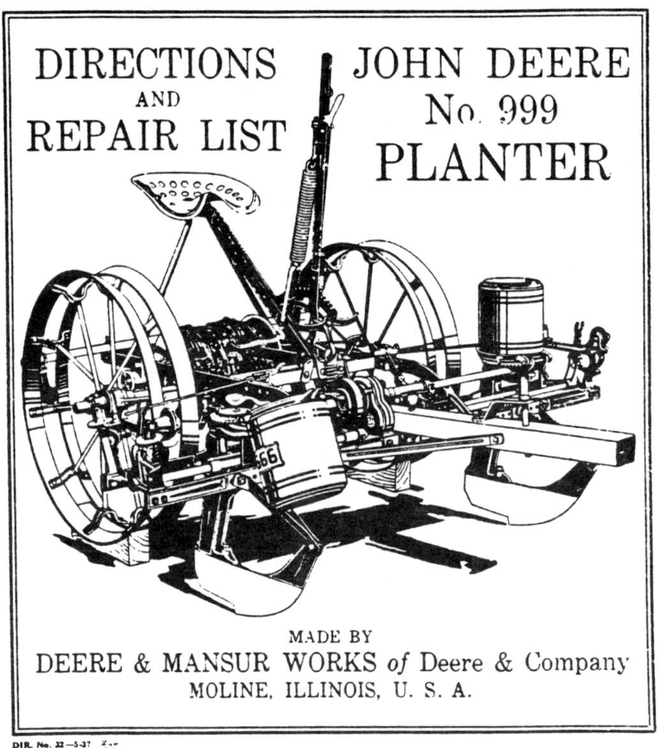

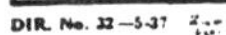

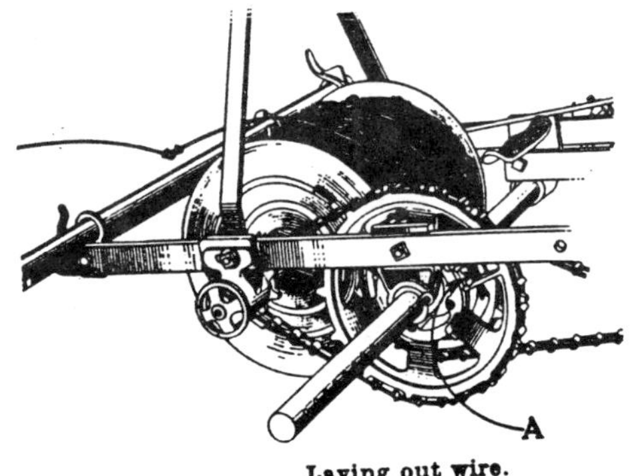

Laying out wire.

Moline, Illinois, 1937

TO OPERATE THE PLANTER IN THE FIELD

When using 28- or 30-inch check-rower wire, a special drive sprocket must be used. This is a divided sprocket which fits over regular sprocket. The number of this sprocket is Y1961 for low-wheel planters, and two of these Y1961 make up the complete special sprocket. The high-wheel divided sprocket is made up of one Y1962 and one Y1963.

TO LAY OUT THE WIRE

Place planter in position on the side of the field where you wish to begin to plant. Put the reel in place with wire running from the top (see cut No. 1 on next page); fasten the end of the wire to the anchor and stake it down next to the fence, or if possible seven feet from center of planter. (No. 1 stake in illustration on next page.)

The bracket holding right-hand bearing of reel is slotted to permit adjusting tension of reel chain.

Be sure to pull the reel sprocket lock casting out of the notch in collar. This automatically makes the axle bearing act as a stop for the lock casting, and will therefore, hold reel sprocket stationary. (See "A" cut No. 1.) Then with the reel friction loosened up slightly, drive straight across the field as this row is a guide for the succeeding rows. A drop of oil on the reel friction will do no harm, but is not necessary if the hand wheel is loosened sufficiently to allow the reel to turn—but without racing, which would result in snarling the wire.

At the end of row, remove the reel and turn into position for planting (see first position). Then stake down the wire in rear of center of the planter leaving enough ground for four rows across the end. (No. 2 stake in cut.) If, owing to the shape of the field the rows increase in length, enough additional wire must be left at one or both ends to reach the longest row.

Now place the wire in the check head (first position); lower the runners into the ground, set variable drop lever on 2, 3 or 4 as desired and drive a steady, even gait to the other end of the field. Do not try to plant clear to the end of the row unless in an open field where the anchor can be placed some distance beyond the edge. Usually room for four rows is left, enough for once across the field and back again.

When as near the end of the field as indicated above, drop the wire to the ground. Raise the runners out of the ground and turn into position for starting on next row (as shown in the second position); then stake down the wire back of center of planter by moving No. 1 stake over to its new position. In driving across the field, the planter tongue should be kept directly over the furrow made by the marker. The hill of corn should be found about 1 to 1-1/2 inches behind the button when wire is in check head. The longer the wire the farther back of button hill should be deposited to get perfect check. If hill is too far forward, lower front end by raising casting "A" in Figs. 1 or 4. If hill is too far back of button, raise front end by lowering casting "A". When setting stakes always keep the wire at the same tension.

LIFTING SPRING ADJUSTMENT

The tension of lifting spring can be adjusted by means of the long bolt screwed into rear of spring.

If this spring is too stiff to permit lever to be pushed forward easily when putting planter in ground, relieve the tension on spring by unscrewing bolt.

When floating planter adjust this spring tension for proper balance.

TO TAKE UP THE WIRE

(Runners should be out of the ground while reeling up wire.) Put reel in place so that the wire winds on from the bottom when the reel turns ahead. Pass the wire through the wire guide under front end of tongue and wire guide on front frame. Pass it under the axle and attach to the underside of the reel drum. To guide or distribute the wire evenly on the reel, use the anchor stake by hooking it upside down through loop in seat standard and moving it slowly back and forth in front of axle (see cut No. 2). **Put reel drive sprocket in gear by pushing lock casting into one of the three notches (see "B", cut No. 2).** The reel is made to revolve faster than the machine moves along the row. Any desired tension can be placed on the reel by means of the friction hub and spring. If it is desired to increase or decrease the tension, tighten or loosen the tension on the friction hub by means of the hand wheel. Keep the friction hub well oiled.

Moline, Illinois, 1937

JOHN DEERE NO. 999K CORN PLANTER

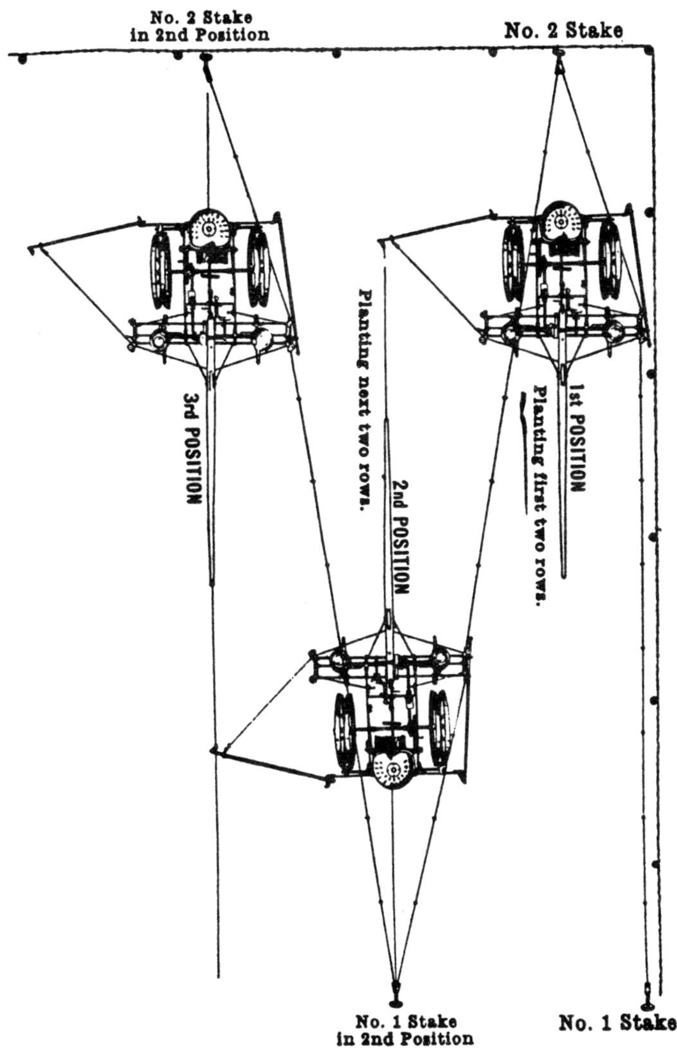

Moline, Illinois, 1937

PLANTERS
AND
RELATED IMPLEMENTS
FROM
1850 TO 1939

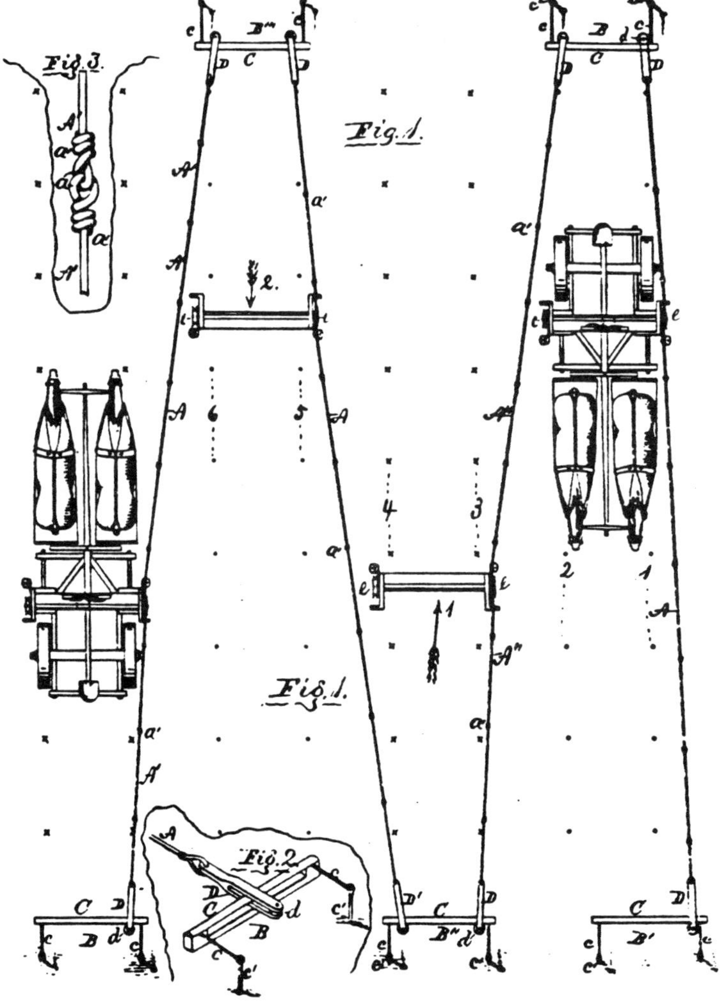

JOSEPH W. FAWKES

December 17, 1850   Grain Drill   7,837

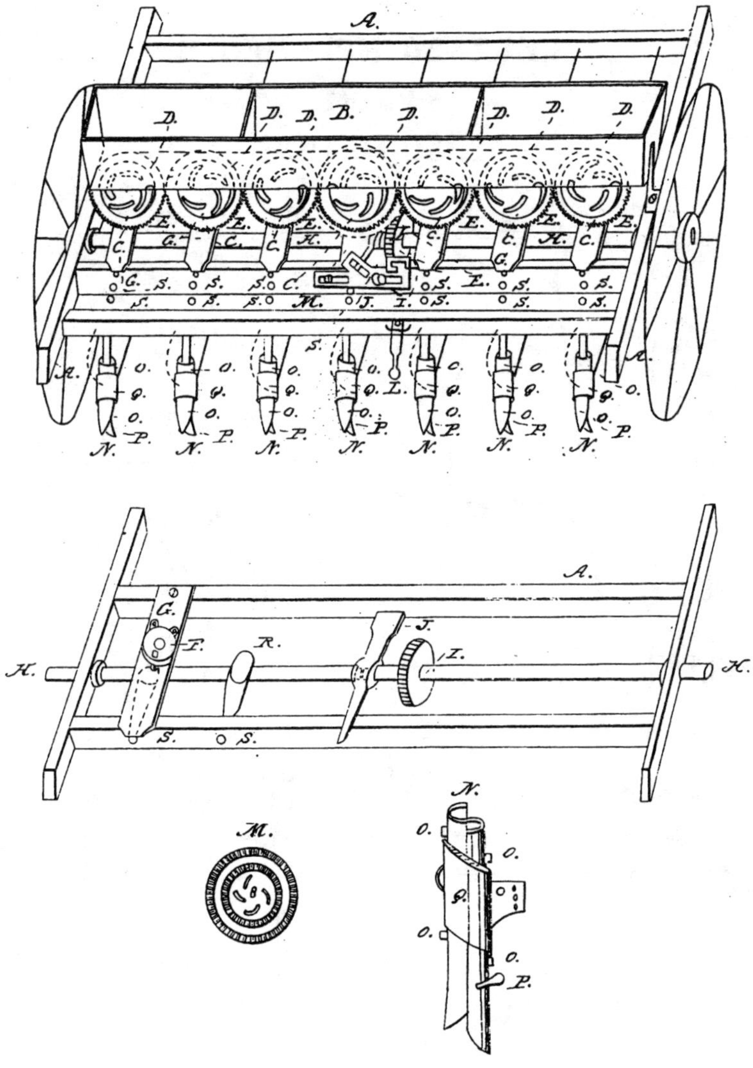

Bart, Pennsylvania

GEORGE W. BROWN

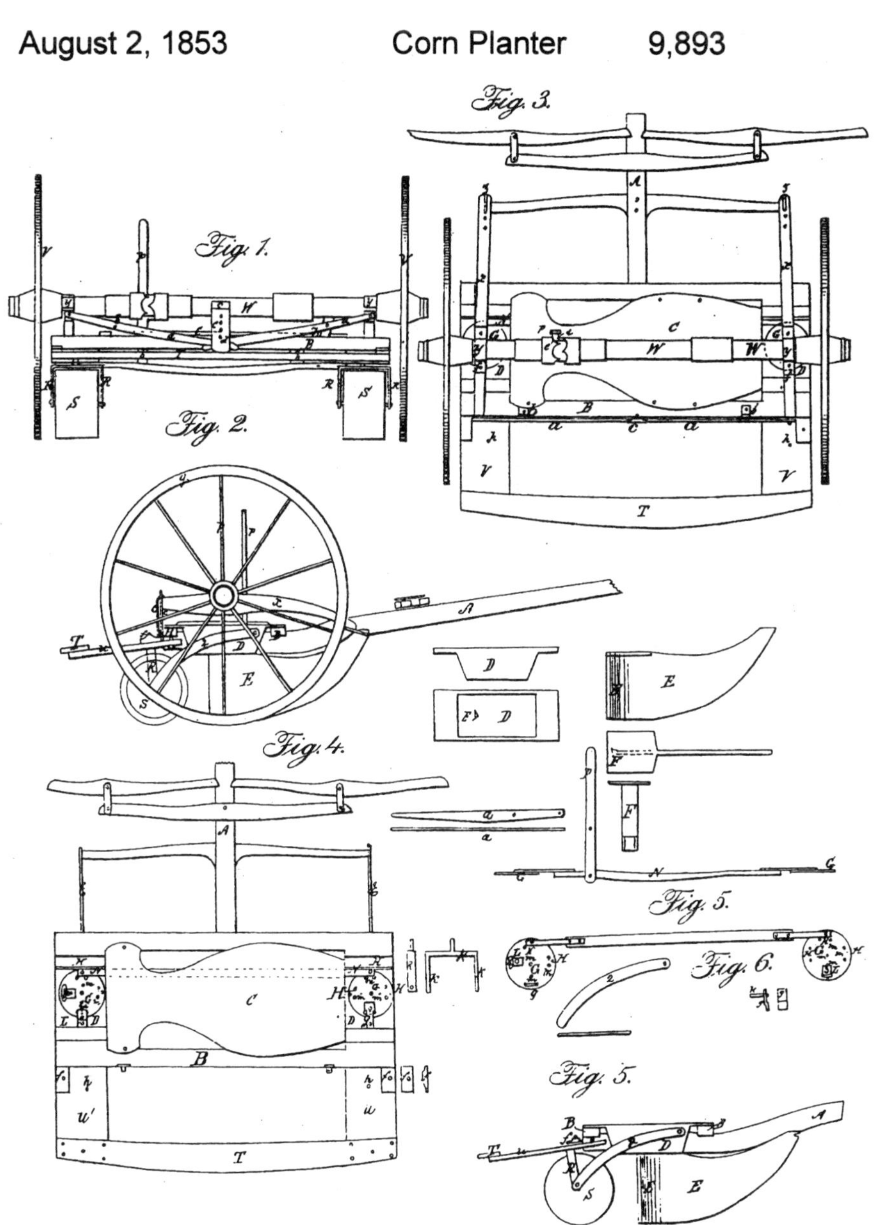

Tylerville, Illinois

January 16, 1855 Seed Planter 12,231

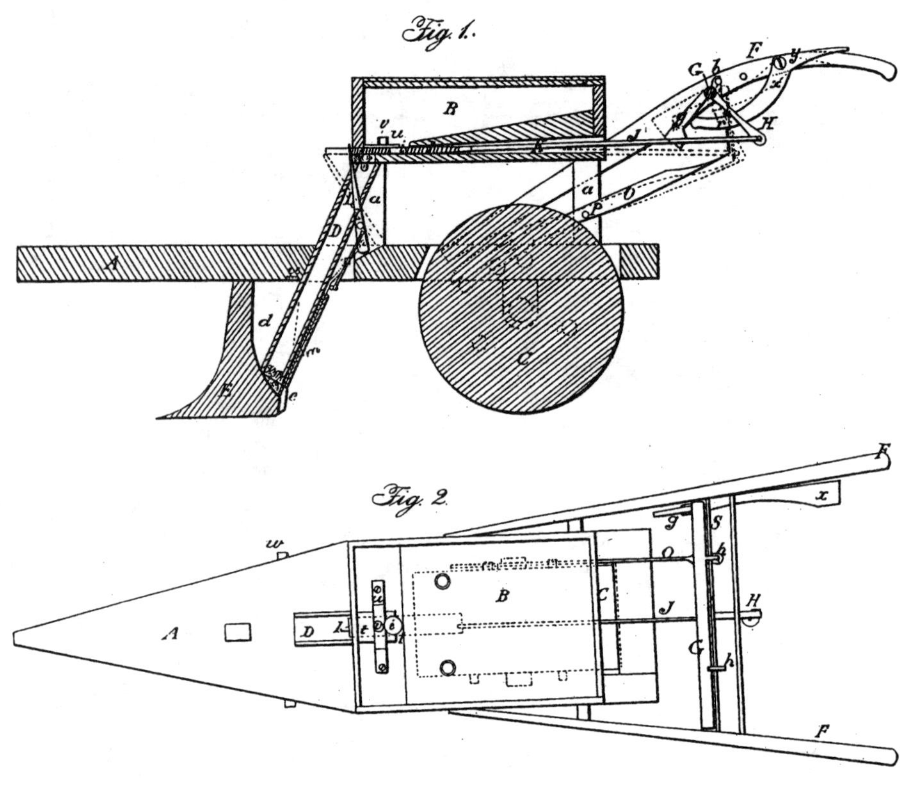

Springfield, Ohio

GEORGE W. BROWN

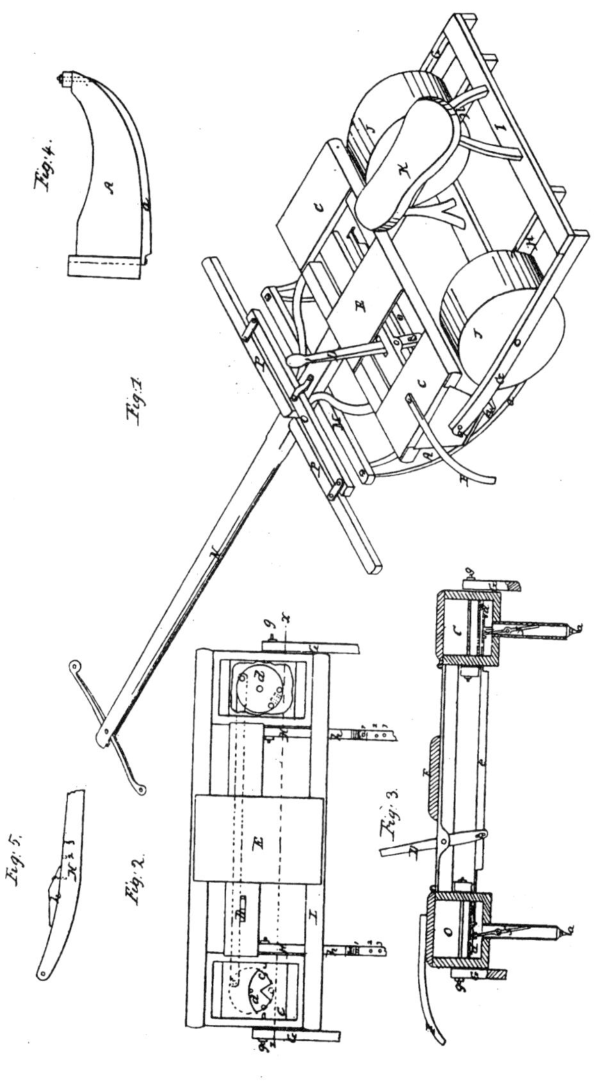

Galesburg, Illinois

EDWARD P. LACEY

April 8, 1856   Corn Planter   14,631

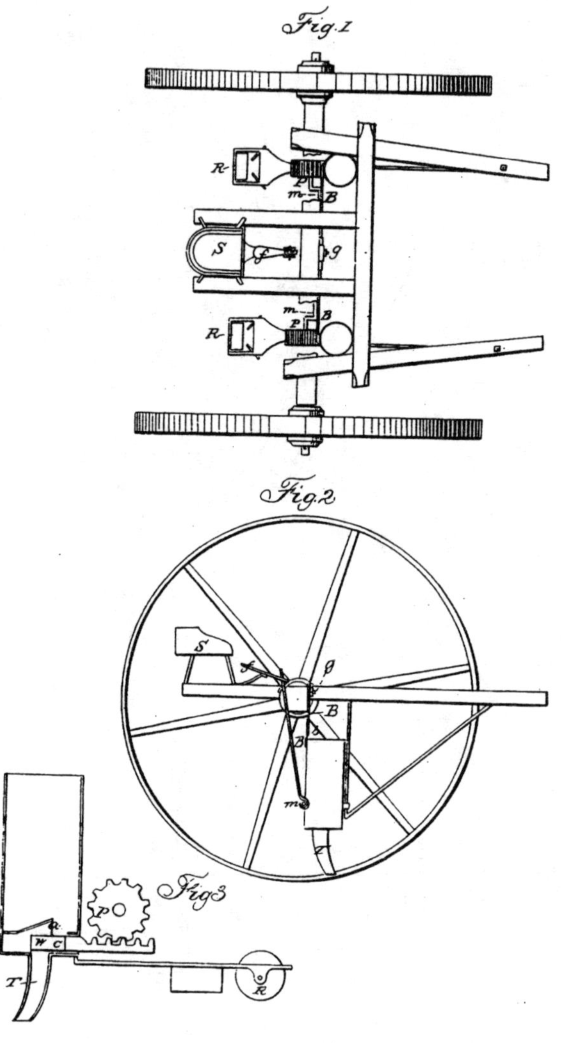

Rochester, New York

December 1, 1857   Corn Planter   18,768

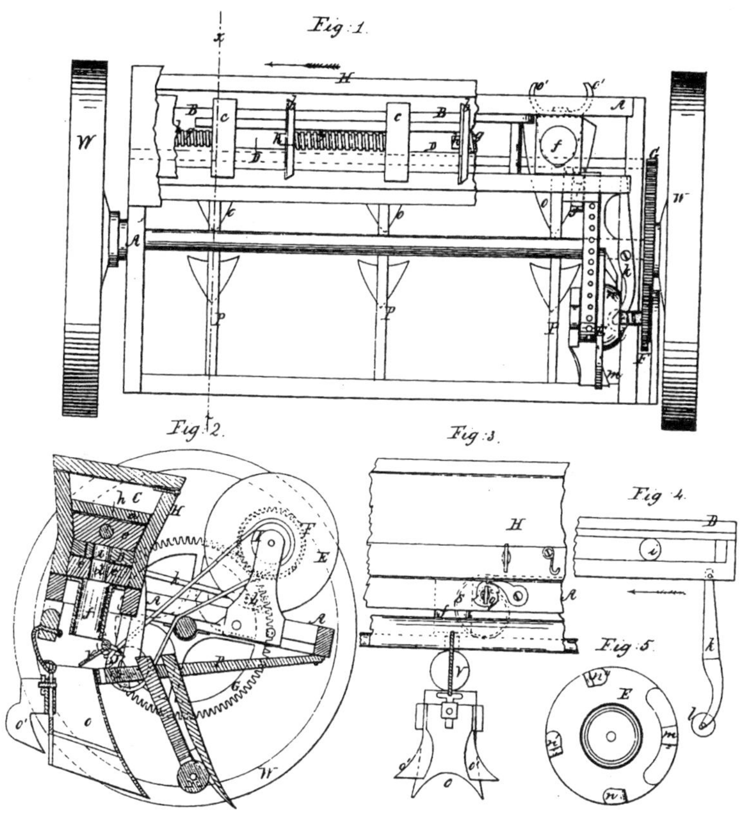

Hamilton, Ohio

May 18, 1858   Seed Planter   20,297

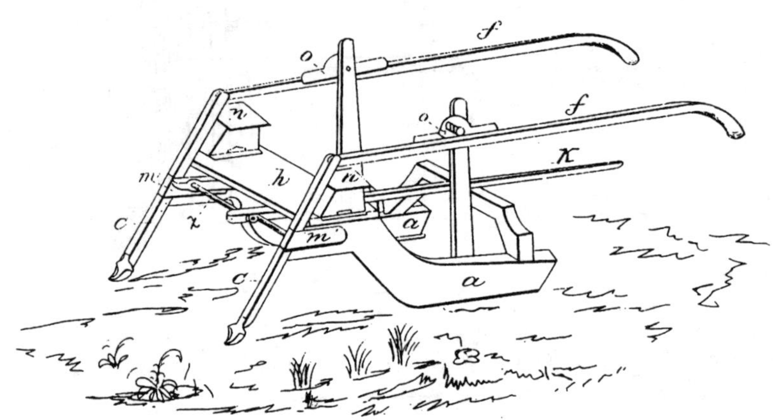

Geneva, Ohio

JACOB HAYNES

March 29, 1859   Corn Planter   23,371

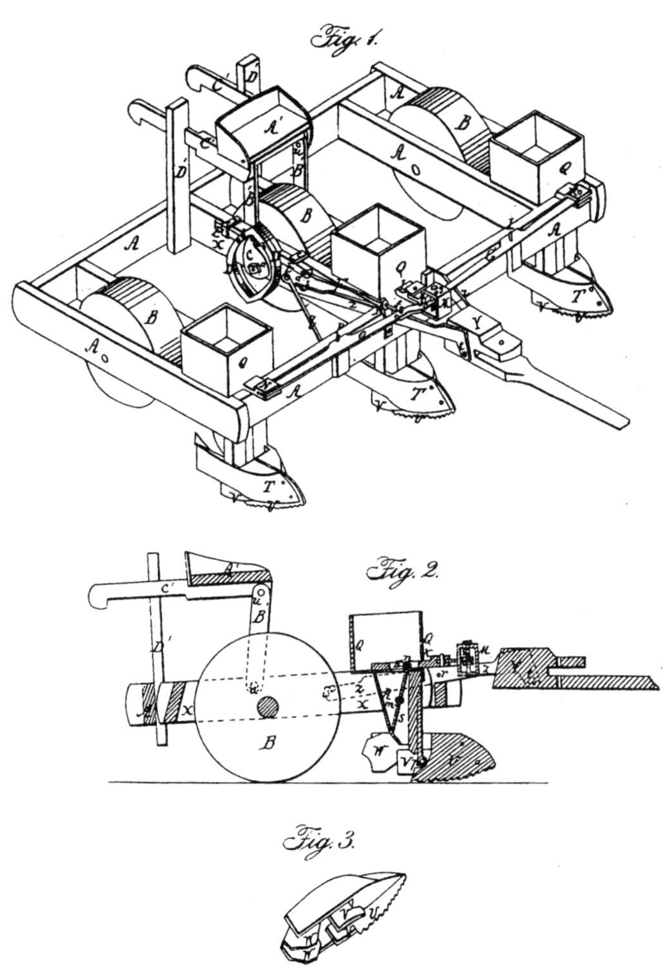

Cameron, Illinois

DAVIS DUTCHER

June 26, 1860 — Corn Planter — 28,847

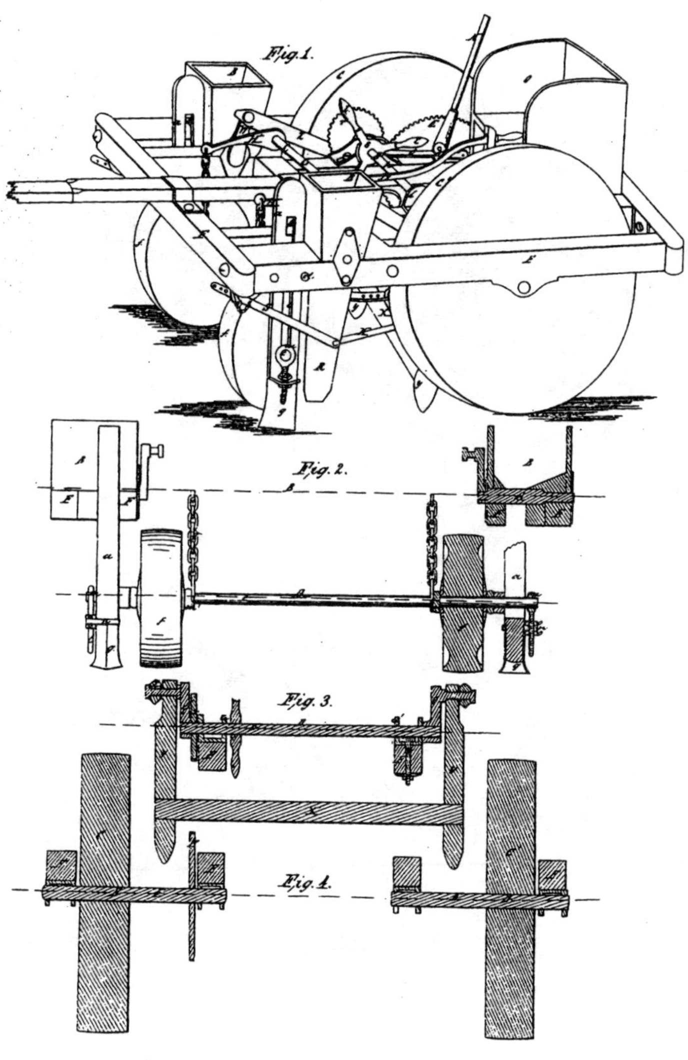

Blue Grass, Iowa

March 19, 1861   Corn Planter   31,700

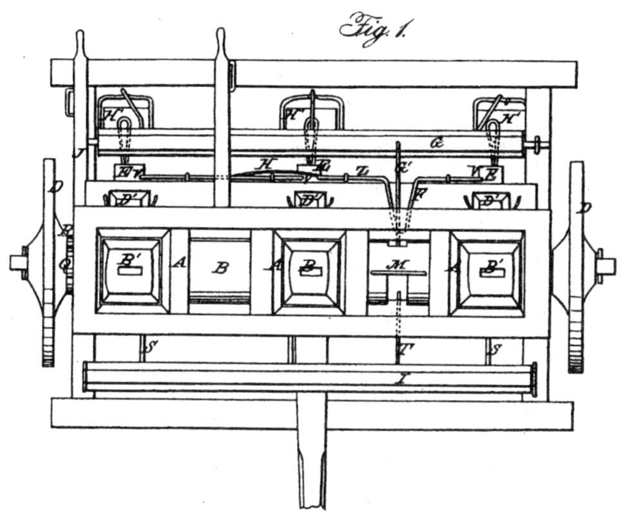

Fig. 1.

Fig 2.

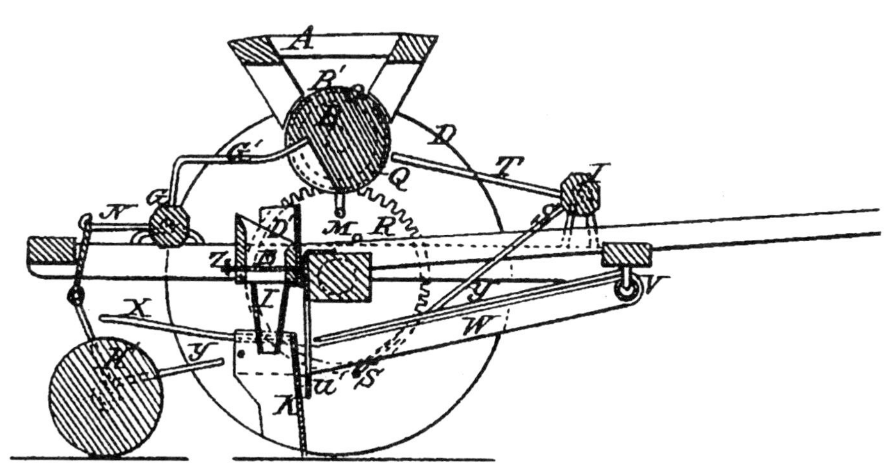

Tafton, Wisconsin

September 3, 1861   33,196

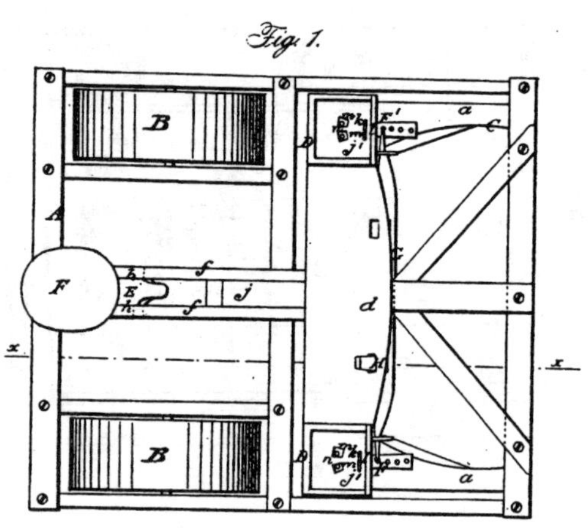

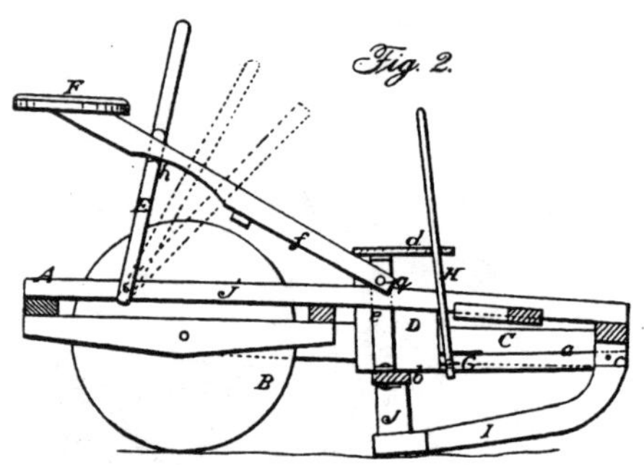

Decatur, Macon County, Illinois

This patent is not a check-row patent. It is included here for its historic value as being the first Haworth Corn Planter patent.

October 18, 1864   44,725

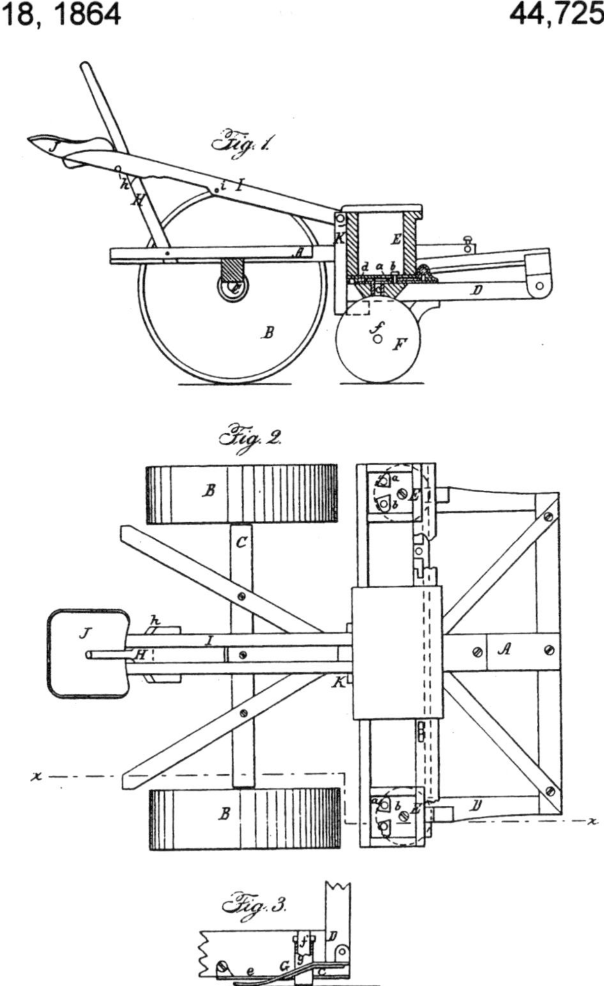

Springfield, Sangamon County, Illinois

This patent is not a check-row patent. It is included here for its historic value as being the second Haworth Corn Planter design that became the base for adapting his first check-row attachment per 100,032 in 1870.

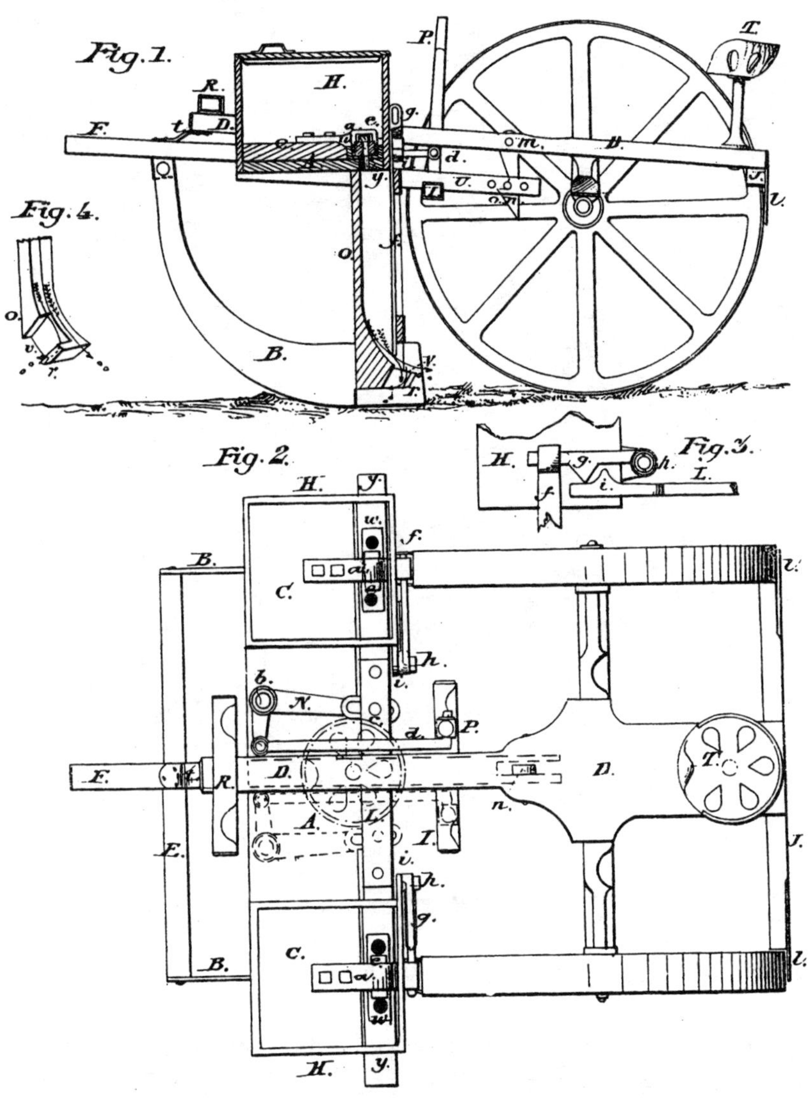

Lafayette, Indiana

DANIEL F. TAFT

October 20, 1868   Corn Planter   83,338

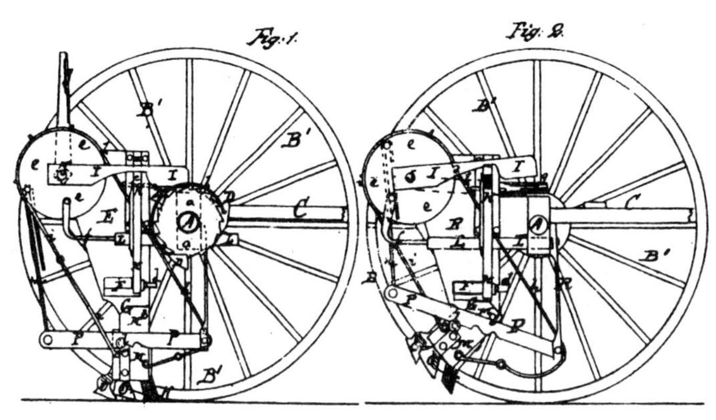

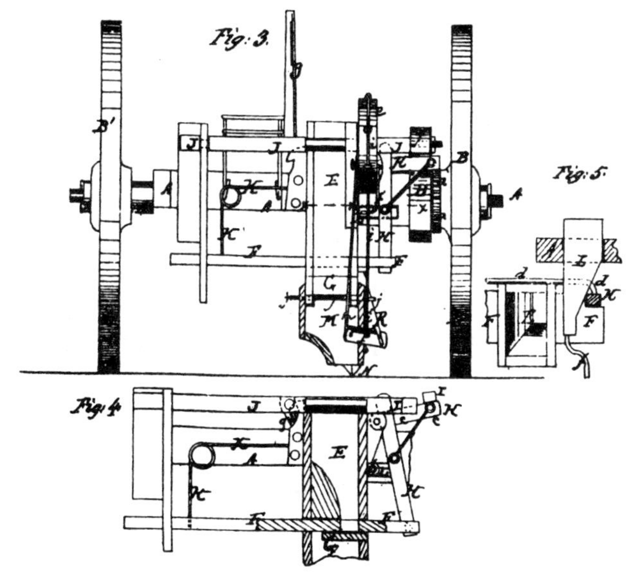

New Bedford, Massachusetts

February 16, 1869   Corn Planter   87,018

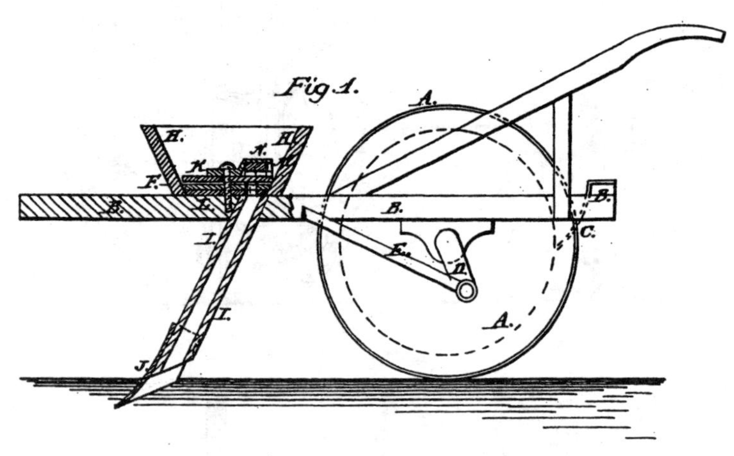

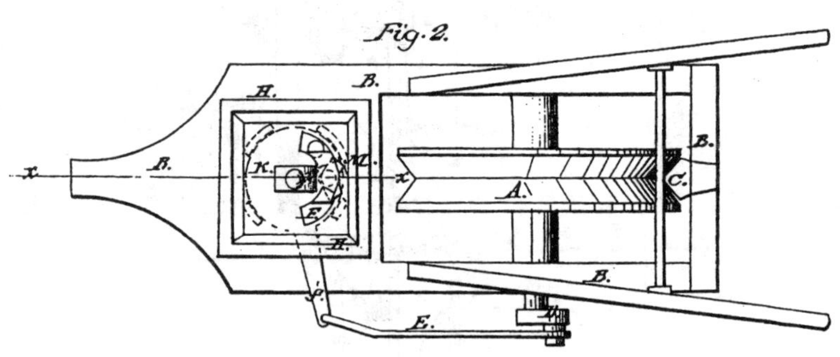

Paris, Illinois

JAMES J. JONES AND SOLOMON H. DWIGHT

July 9, 1872   128,888

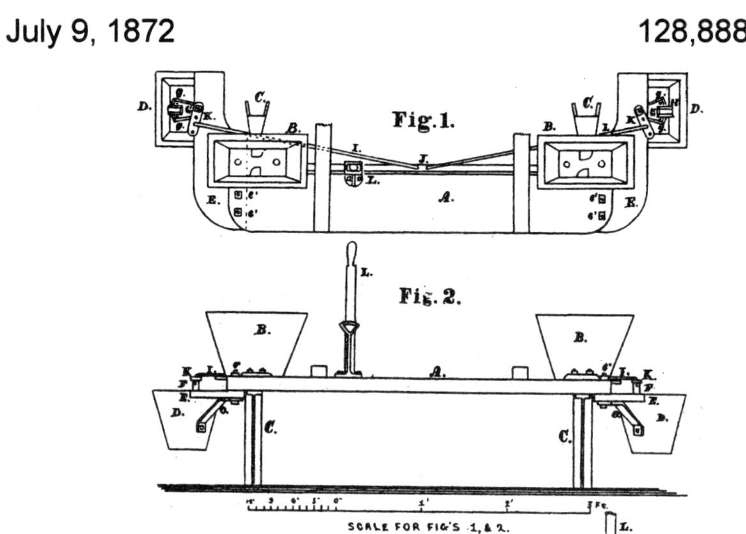

Decatur, Macon County, Illinois

"Our invention relates to an improvement in corn planters, obviating the marking off of the field previous to planting; and consists in the use of boxes attached to the frame work that supports the seed boxes and furrow shares, said boxes to be filled with lime, plaster, dry wood ashes, or some substance different in color from the ground. A suitable device is placed in the box and attached to the connecting bar that operates the seed dropping mechanism, so that when the seed is dropped in the ground a small quantity of lime, plaster, or wood ashes is dropped, leaving a spot or mark on the ground, said spots or marks, serving as guides in planting subsequent rows in check."

"After the first two rows are planted one of the connecting rods can be removed, (at the option of the driver) as a mark on one side of the planter will be sufficient for a guide. This marking attachment can be attached to any corn planter."

CHARLES HUTCHINS

October 21, 1873   Corn Planter   143,826

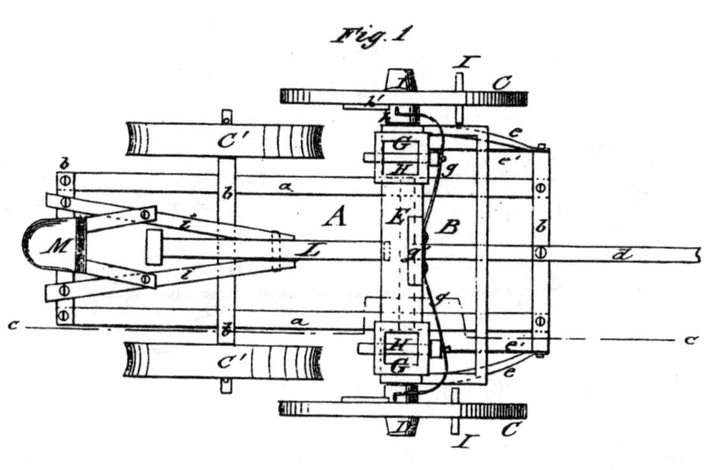

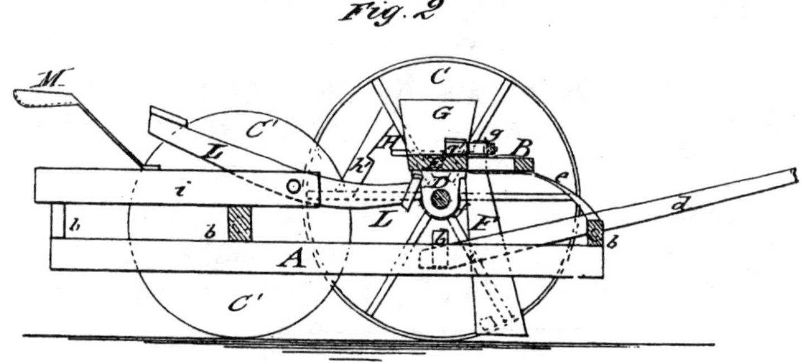

Aubrey, Kansas

G.W. BROWN

June 2, 1874   Corn Planter   151,561

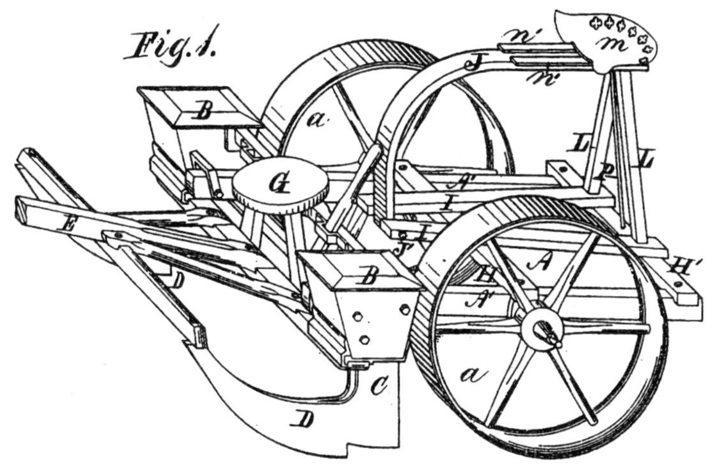

Galesburg, Illinois

June 23, 1874   Corn Planter   152,434

"My invention relates to that class of double-row planters in which the furrows are opened by runners, and the feed-slide operated by a boy riding on the front; . . . ."

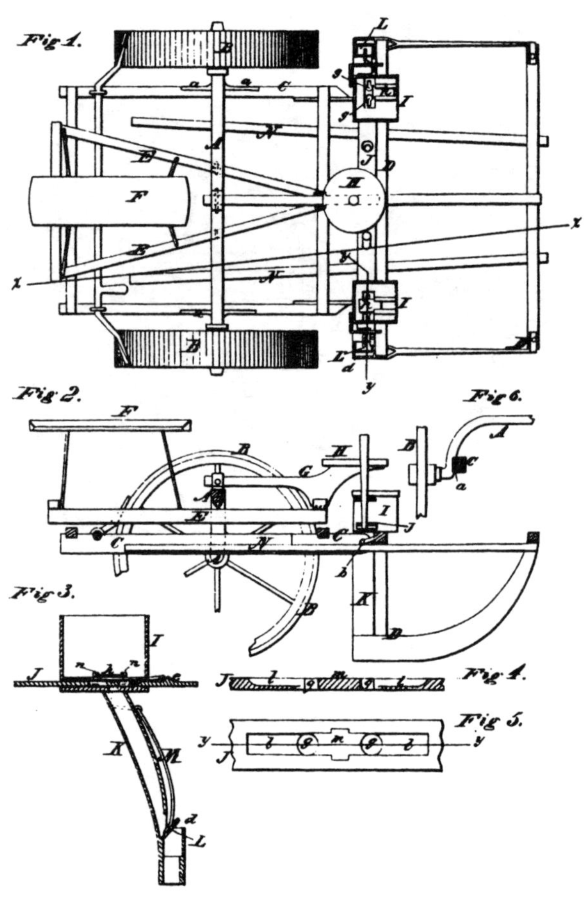

Grand Junction, Iowa

February 15, 1876   C-R Attachment   173,650

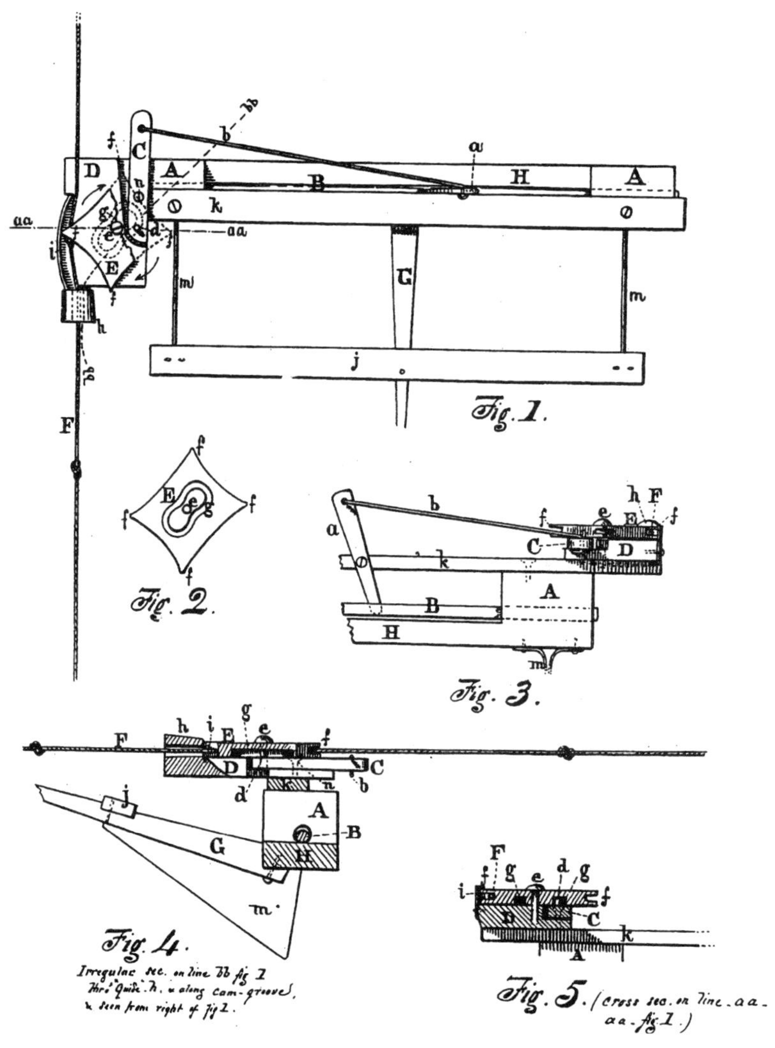

Fairbury, Illinois

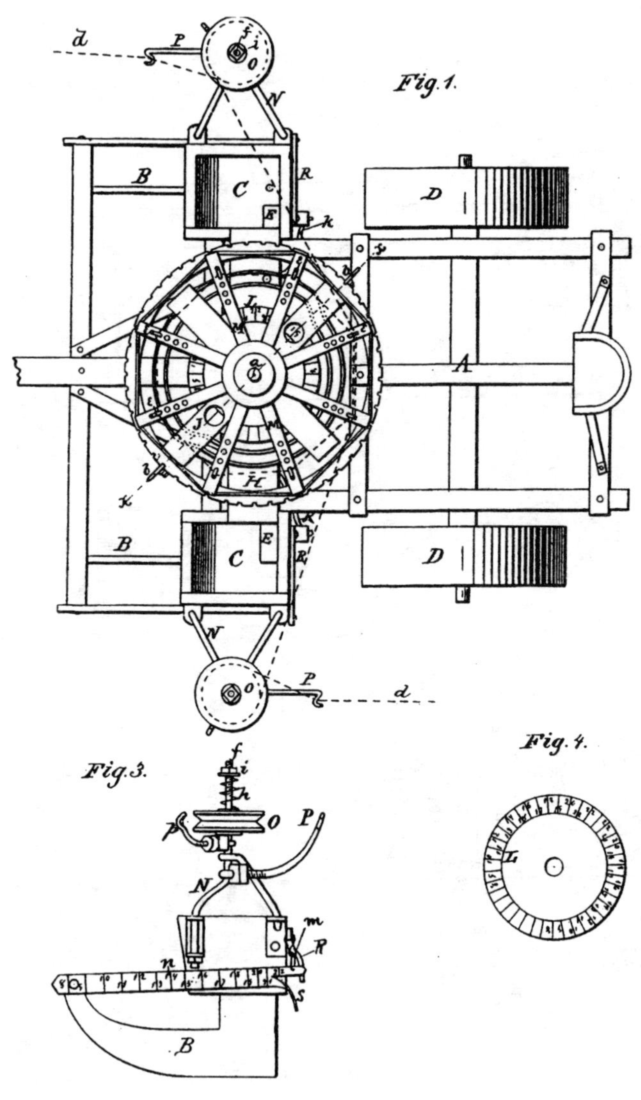

Aledo, Illinois

WILLIAM GILMAN

January 16, 1877   Corn Planter   186,203

Note the check-line path straddled by the team.

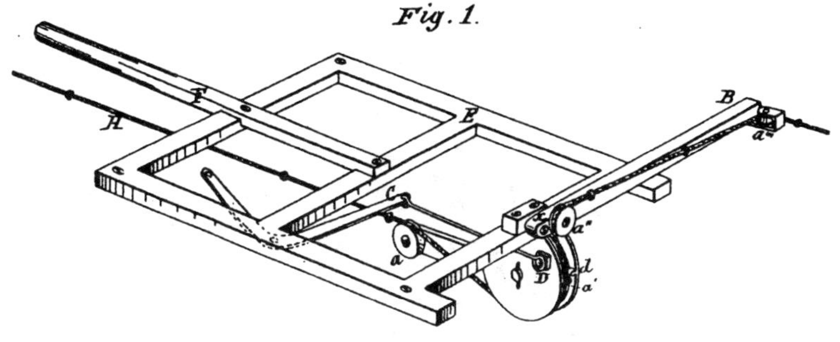

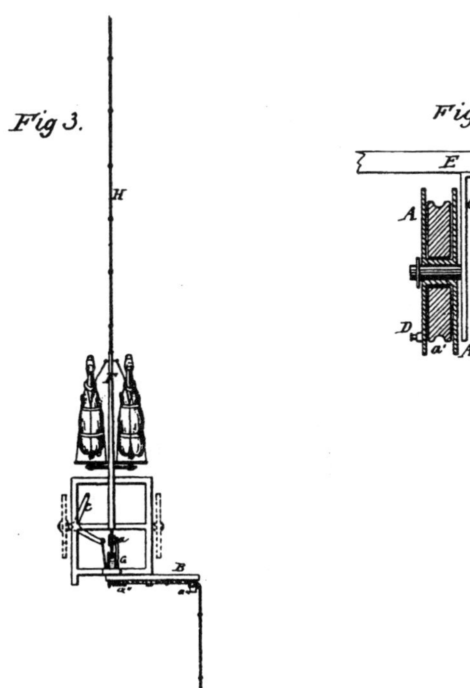

Chicago, Illinois

February 20, 1877   Check-Rower   187,610

Note that the knot appears to be the Barnes re-issue 7,522.

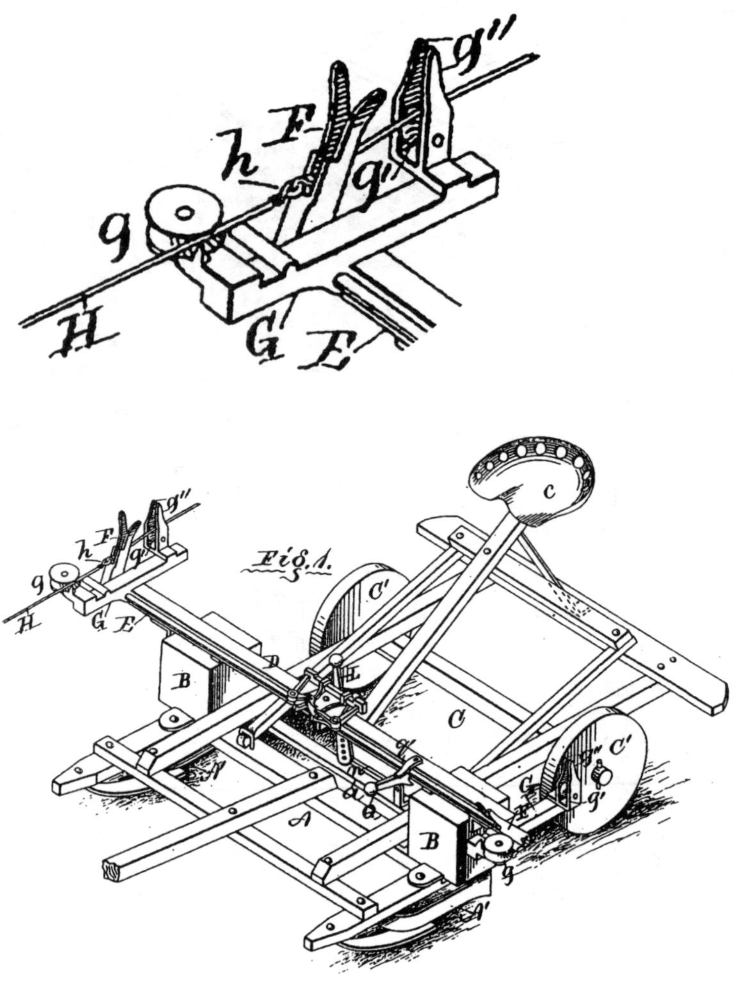

Decatur, Illinois

JAMES W. PERRY

March 27, 1877   Self-Dropper   188,942

An early mechanical driven seed dropper.

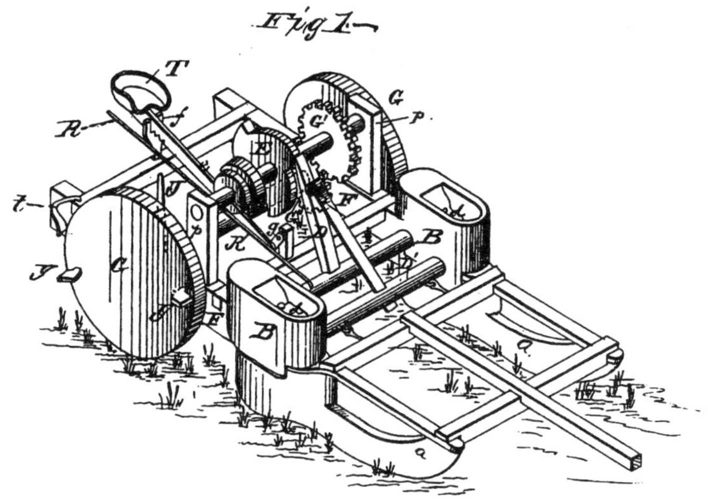

Keytesville, Missouri

April 17, 1877   Drop Marker   189,581

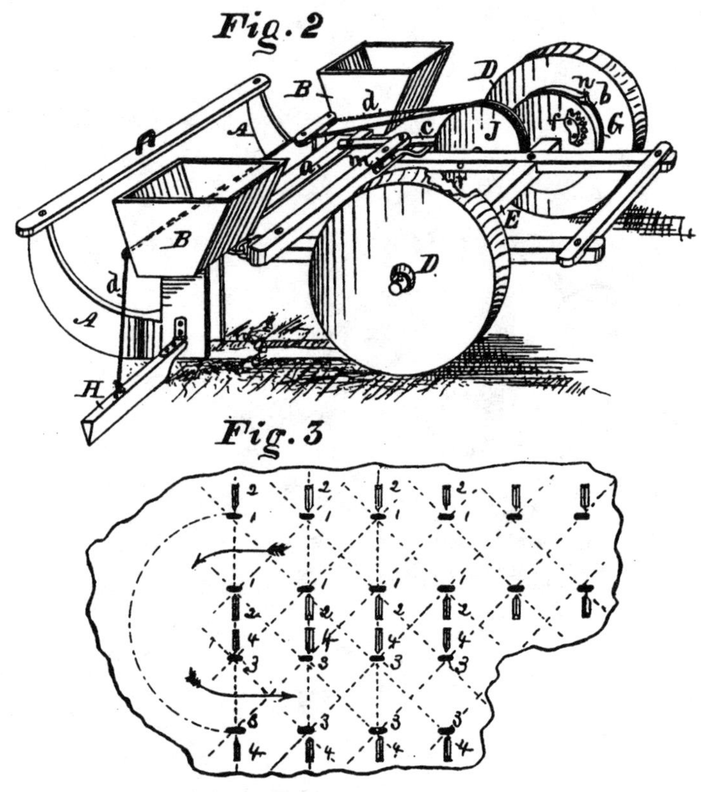

Little Sioux, Harrison County, Iowa

This invention provides for a wood marker, rope controlled and timed with the seed drop, to plot the next row.

October 9, 1877   C-R Attachment   196,012

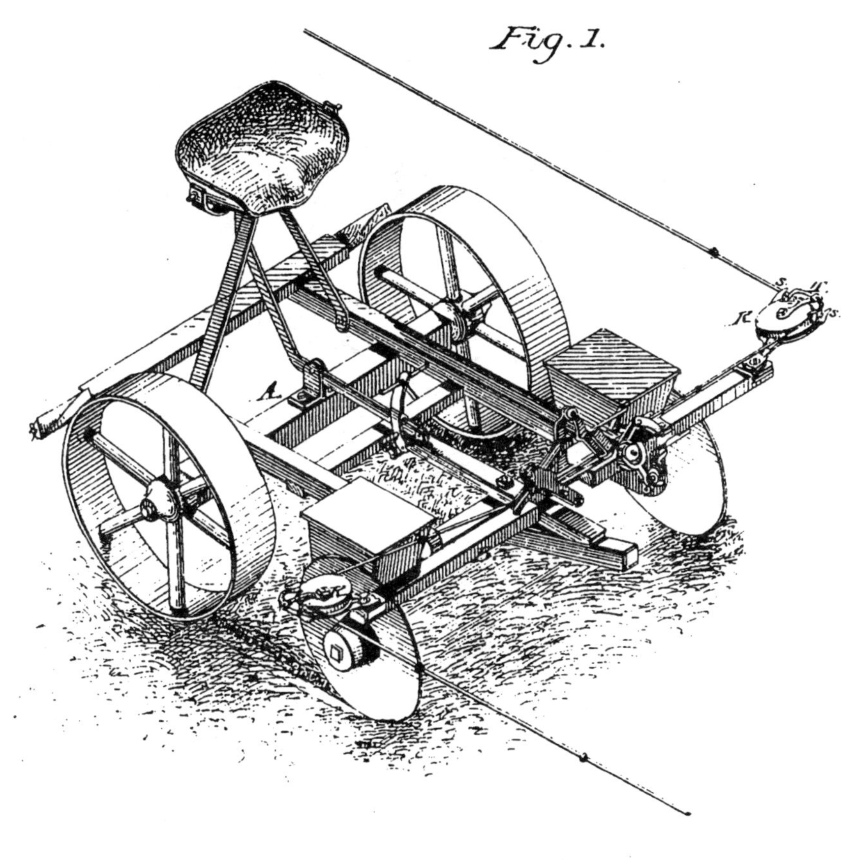

Decatur, Illinois

November 20, 1877   Hand C-R Planter   197,272

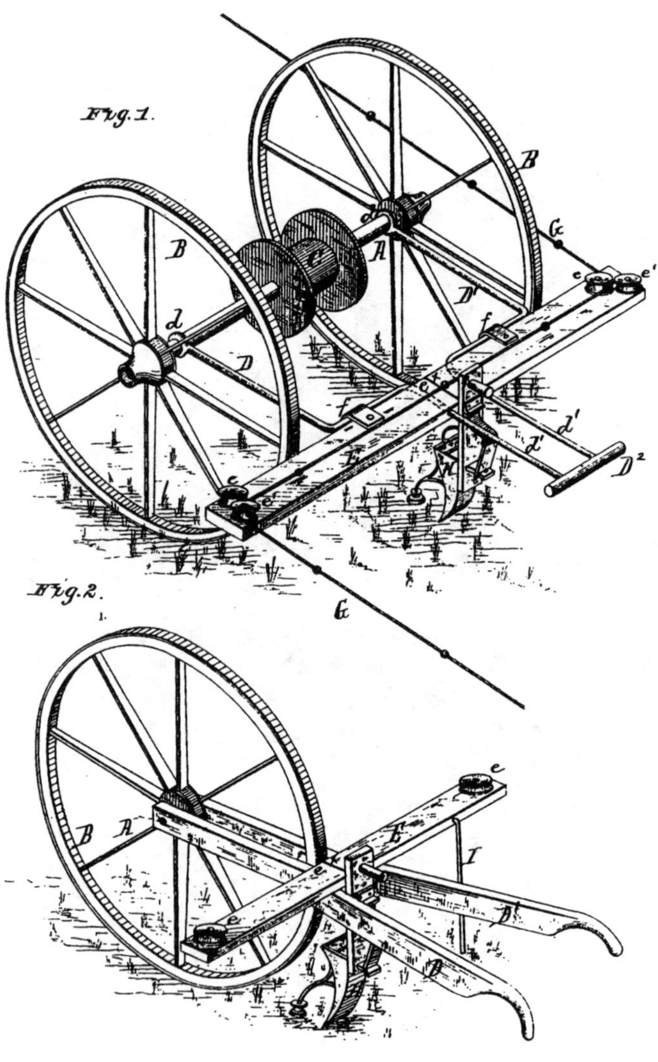

Decatur, Illinois

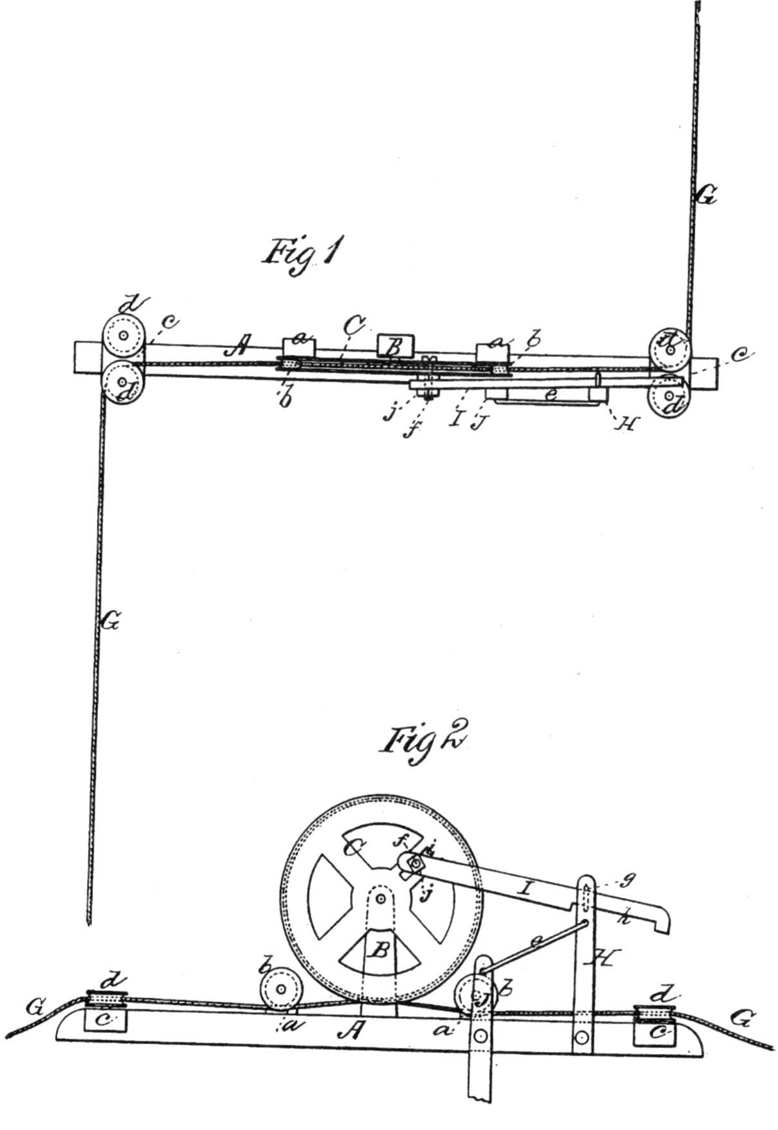

Decatur, Illinois

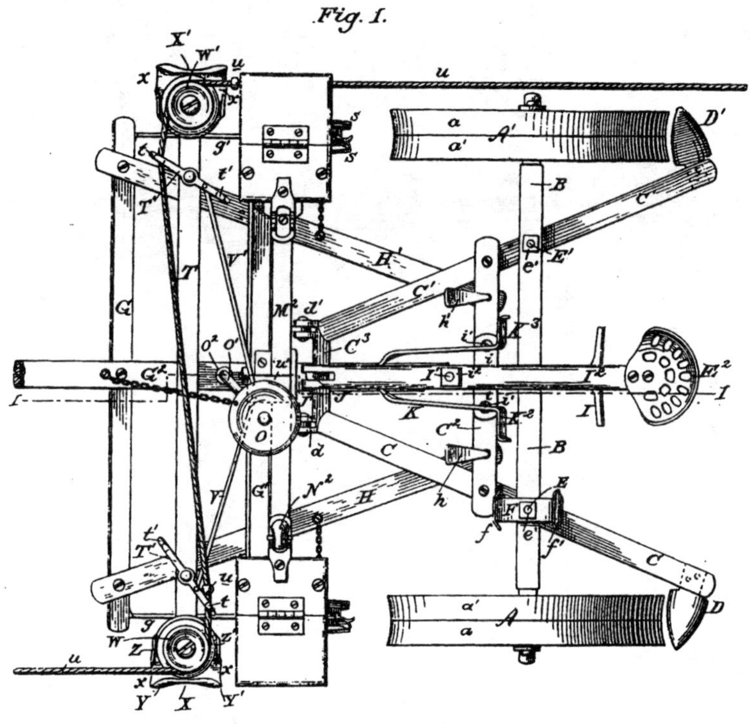

Rockford, Illinois

TYLER C. LORD

November 26, 1878   C-R Attachment   210,340

This uses a knotless rope as a friction drive, one of many such patents.

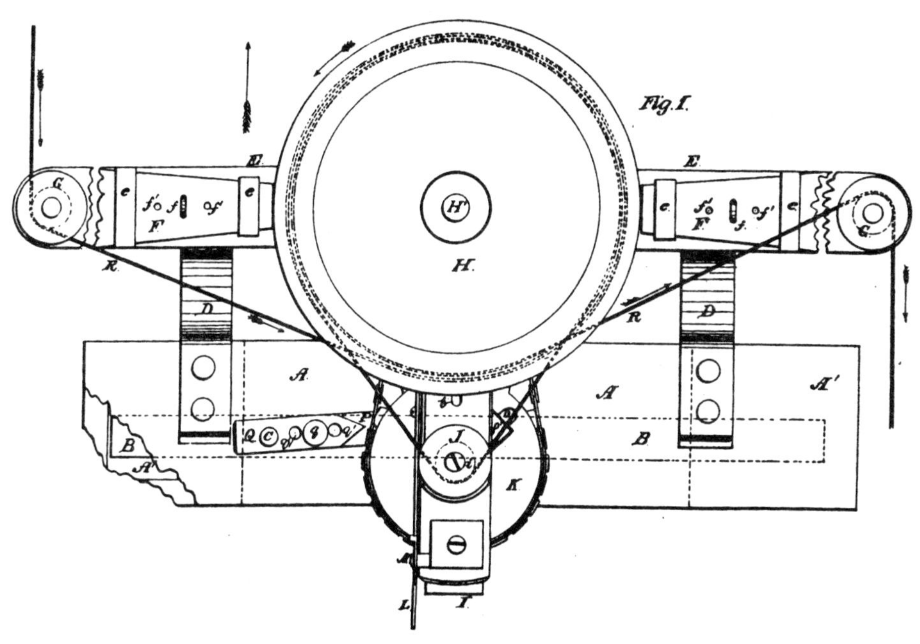

Joliet, Illinois

JAMES BABCOCK

April 8, 1879   Check-Rower and Drop   214,079

This uses a knotless rope as a friction drive.

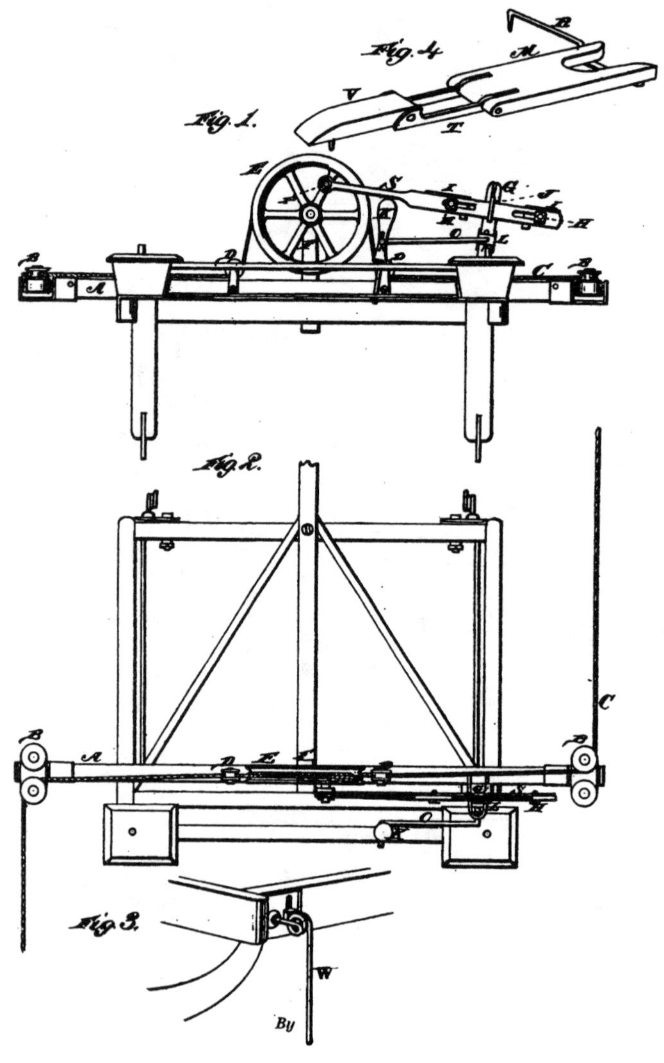

Decatur, Illinois

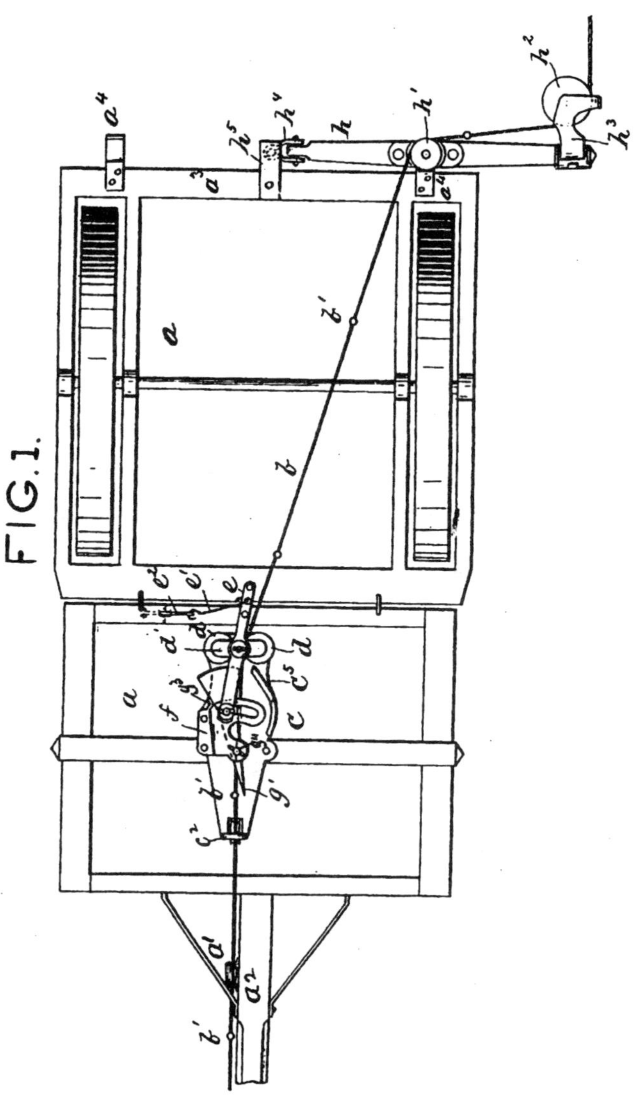

Tuscola, Illinois

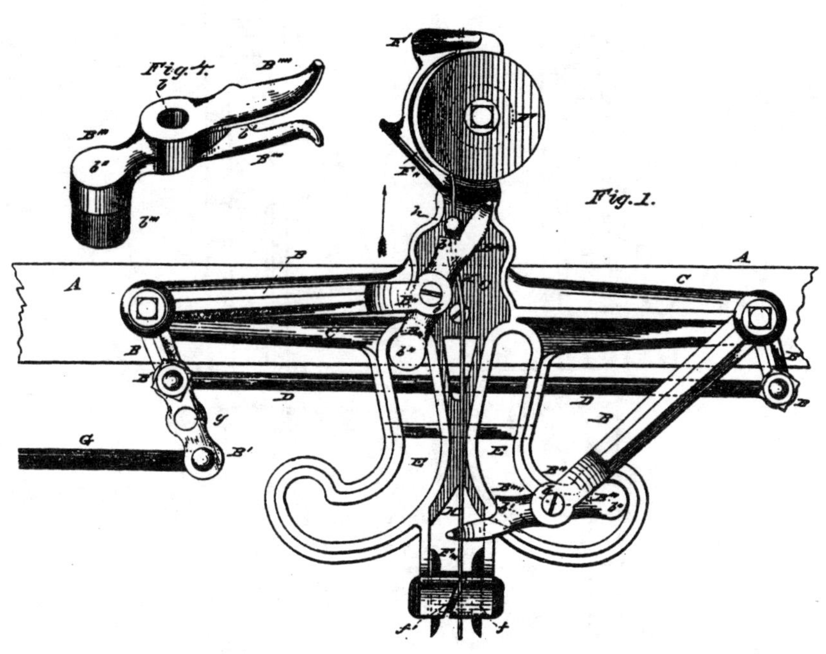

Galesburg, Illinois

NATHAN C. BOLIN

July 25, 1882   Set Out/Take Up Unit   261,658

This was also used to set out and take up barbed wire.

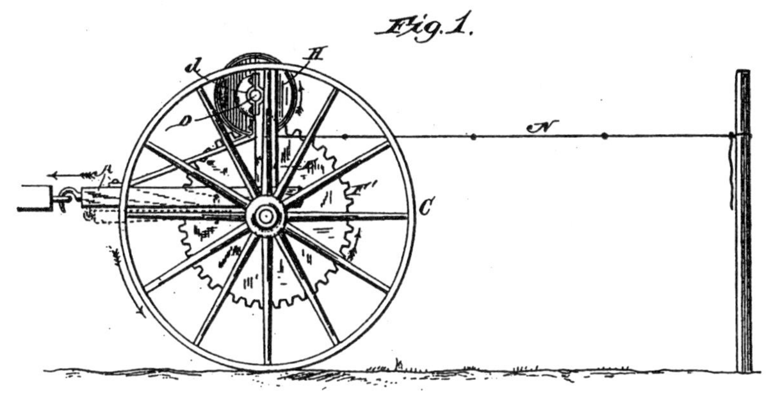

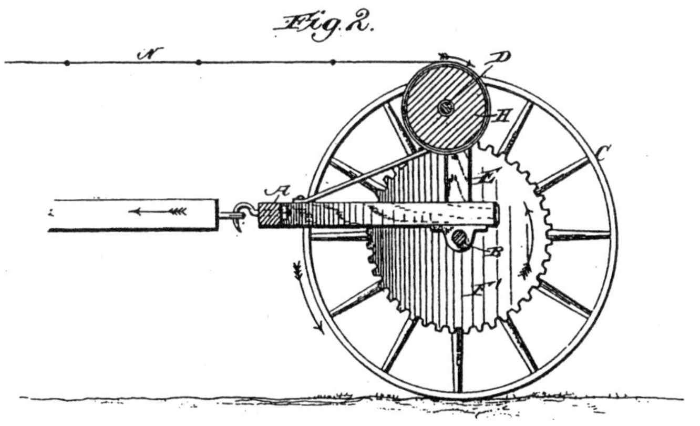

Red Oak, Iowa

October 24, 1882 C-R Attachment 266,341

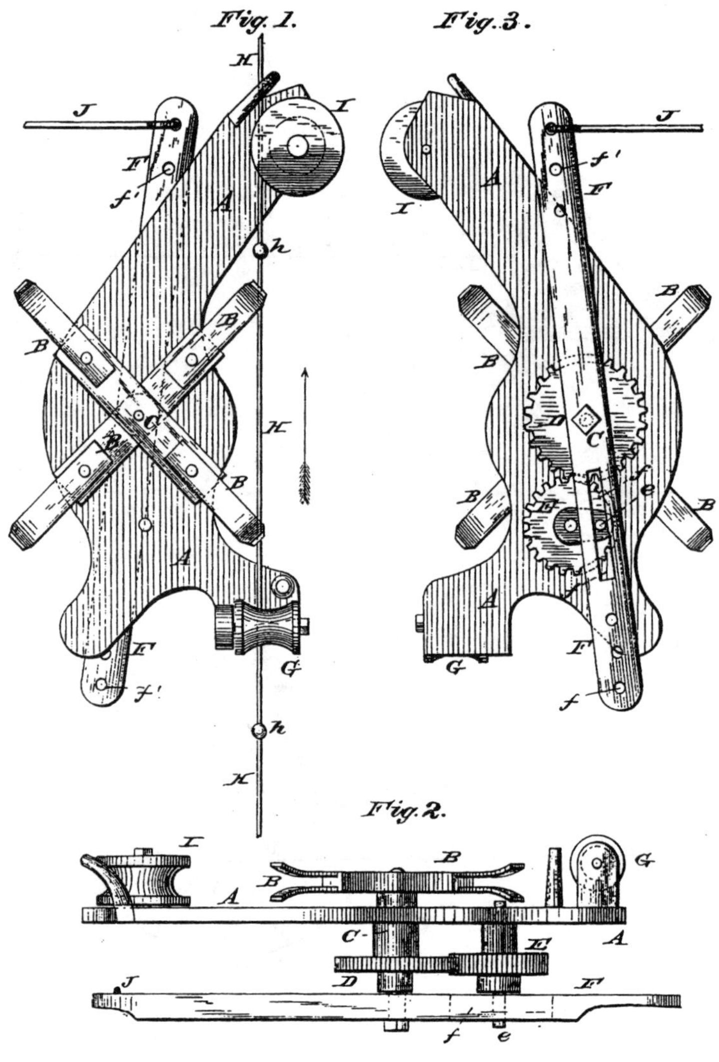

Galesburg, Illinois

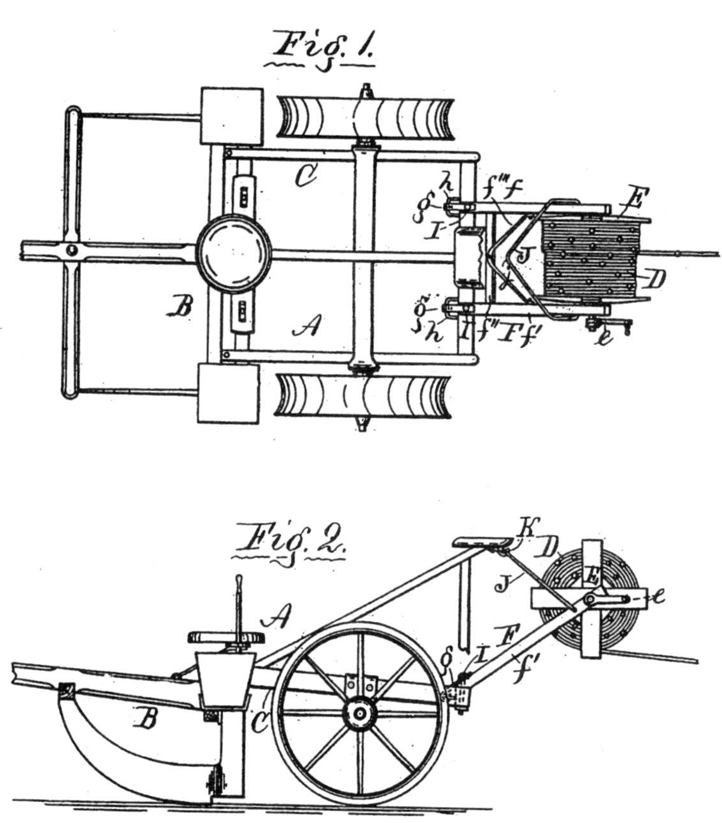

Alpha, Illinois

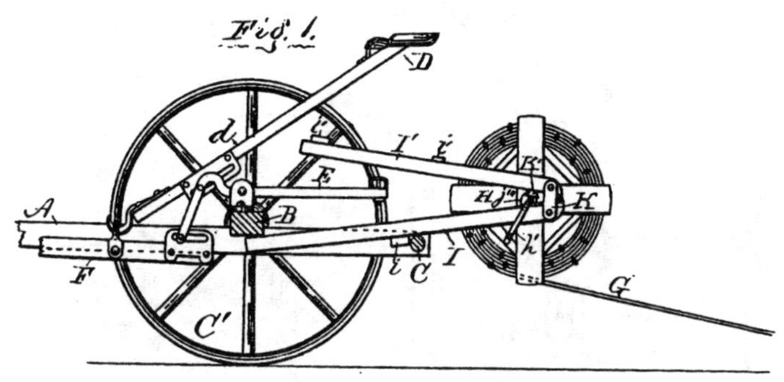

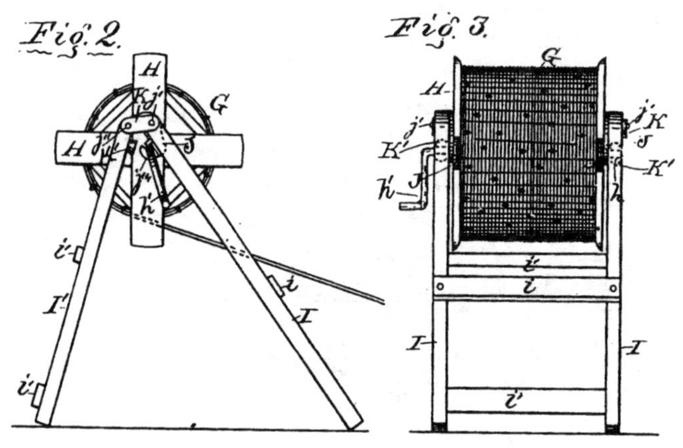

Galesburg, Illinois

July 24, 1883   C-R Attachment   281,919

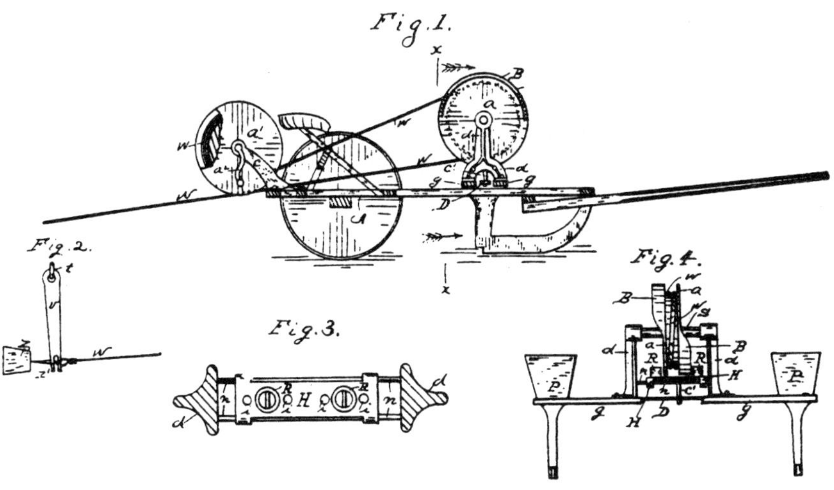

Joliet, Illinois

February 26, 1884   C-R Planter   294,269

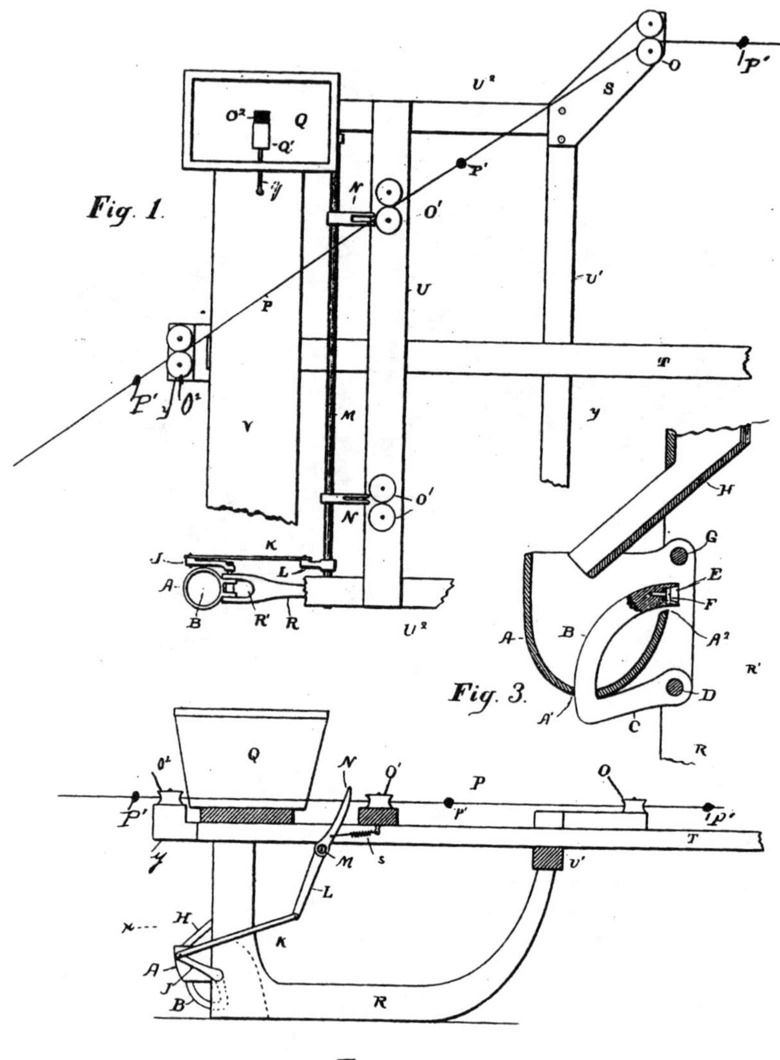

Peoria, Illinois

April 7, 1885   Planter Check-Rower   315,226

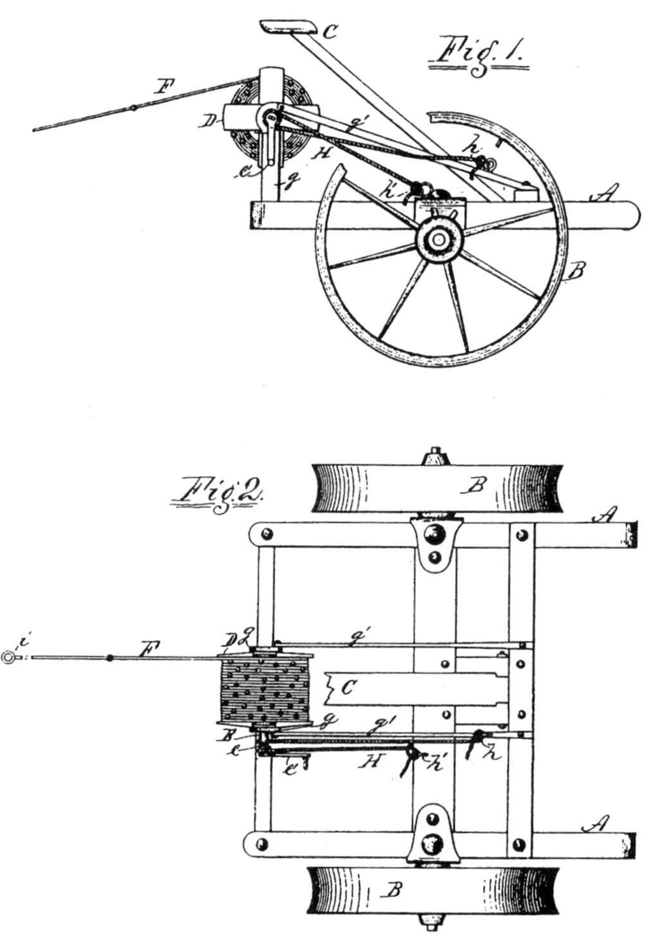

Decatur, Illinois

April 7, 1885   Planter Check-Rower   315,446

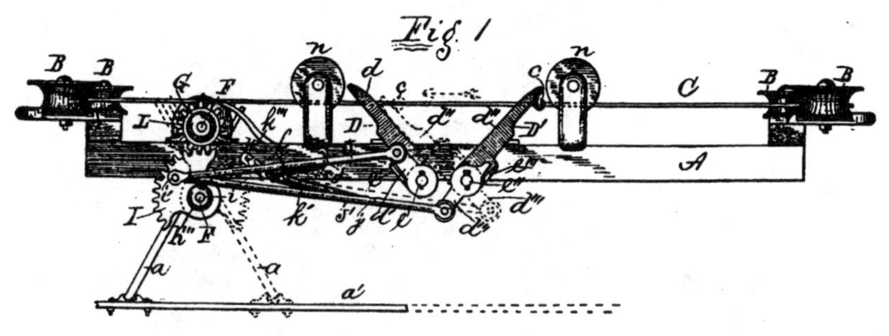

Fig. 1

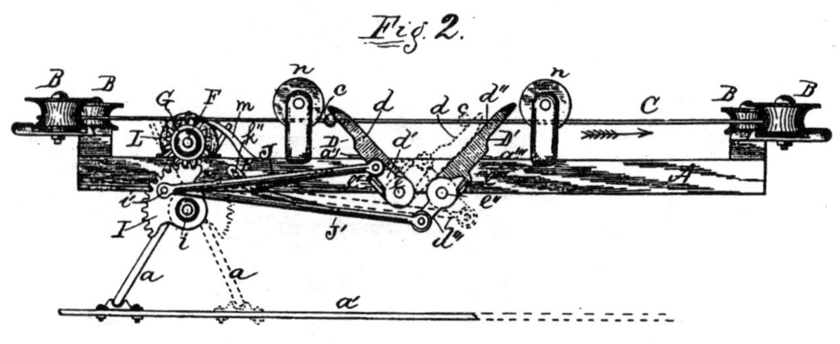

Fig. 2.

Decatur, Illinois

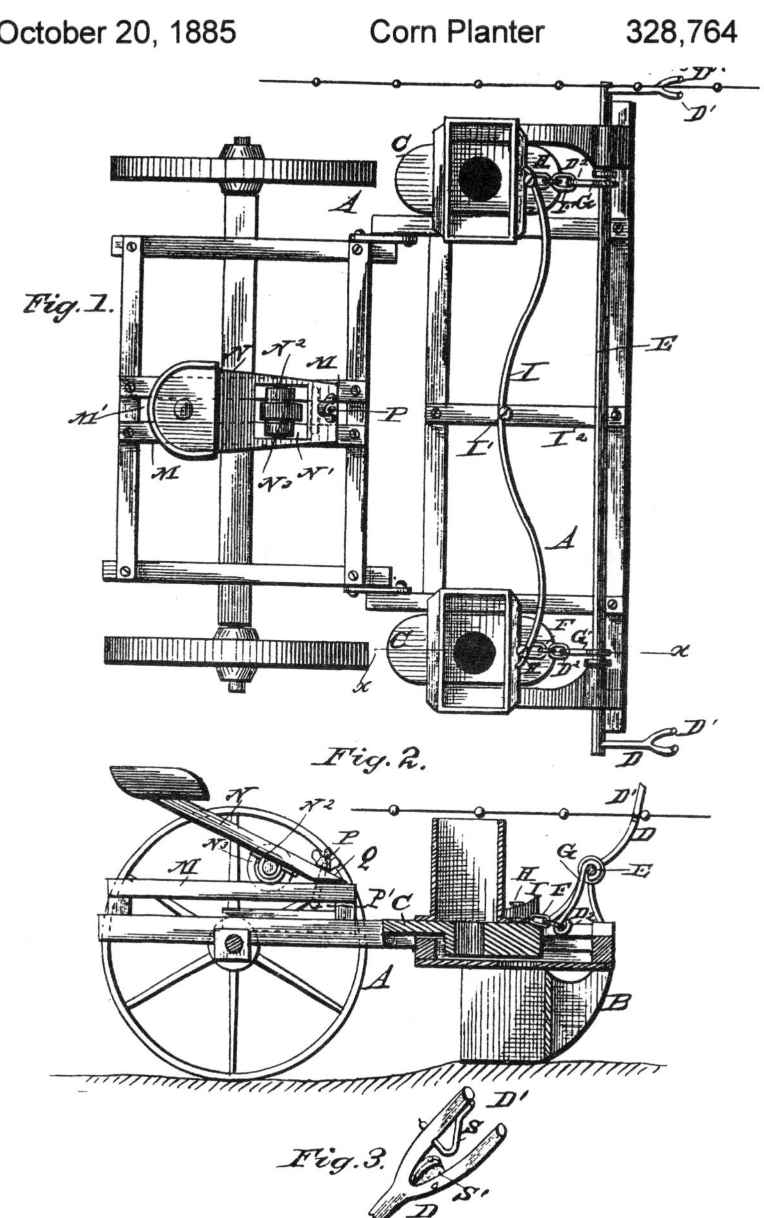

Buffalo, Illinois

Shelbyville, Illinois

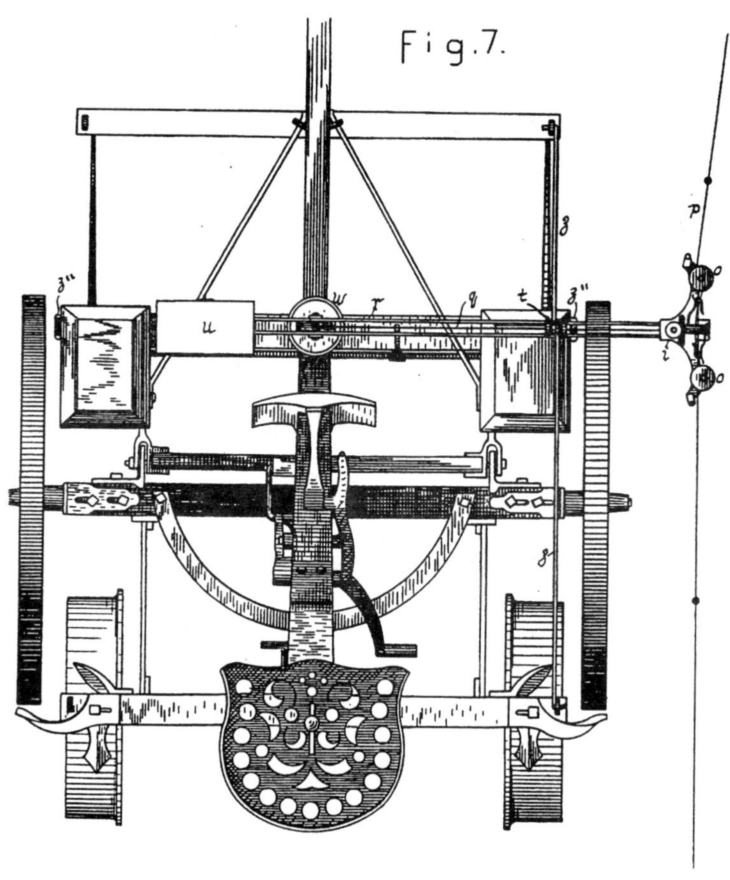

Decatur, Illinois

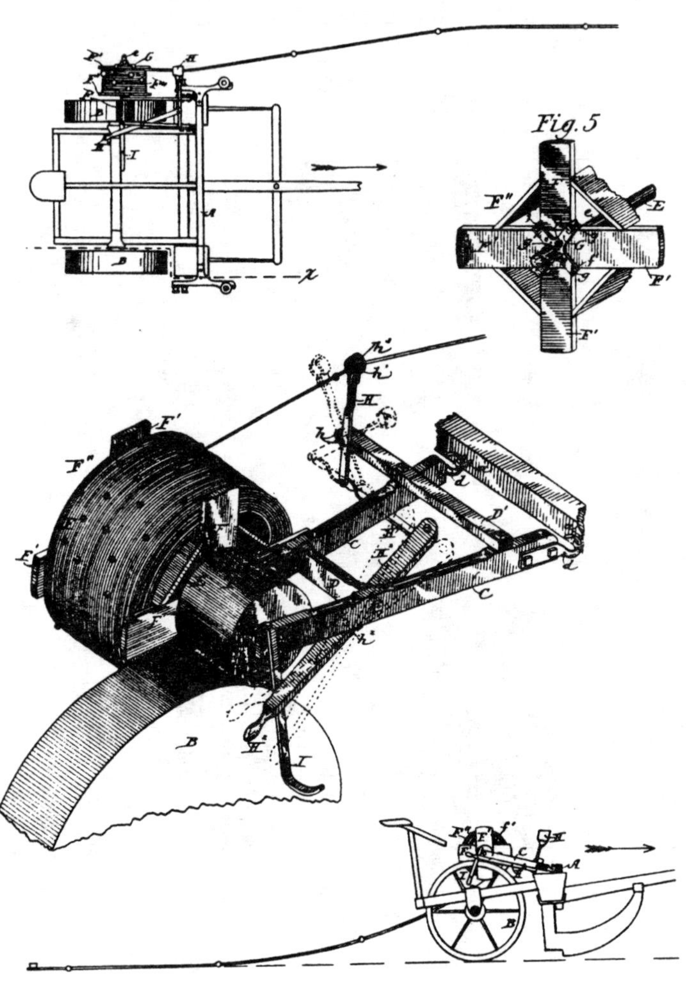

Galva, Illinois

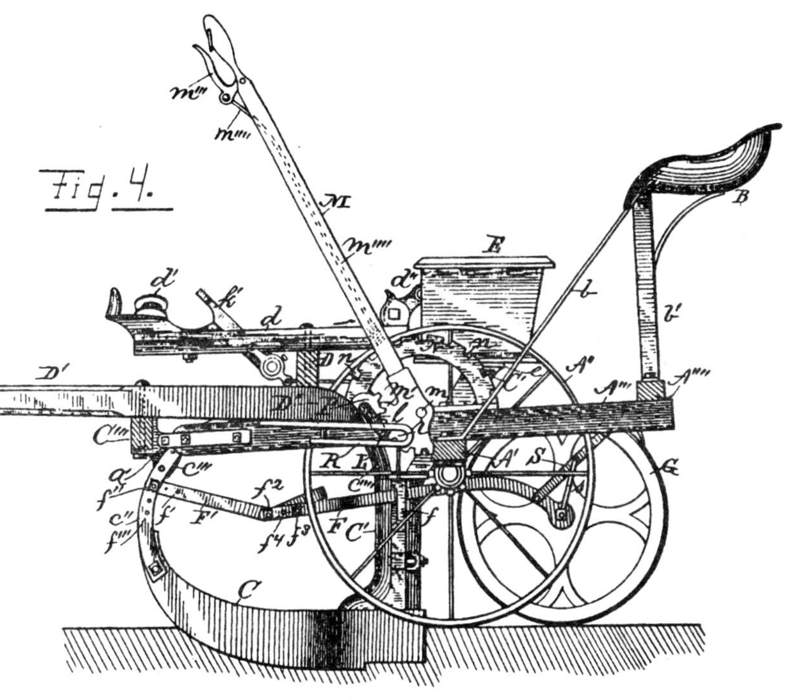

Decatur, Illinois

LORENZO D. BENNER

August 21, 1888   Check-Rower   388,013

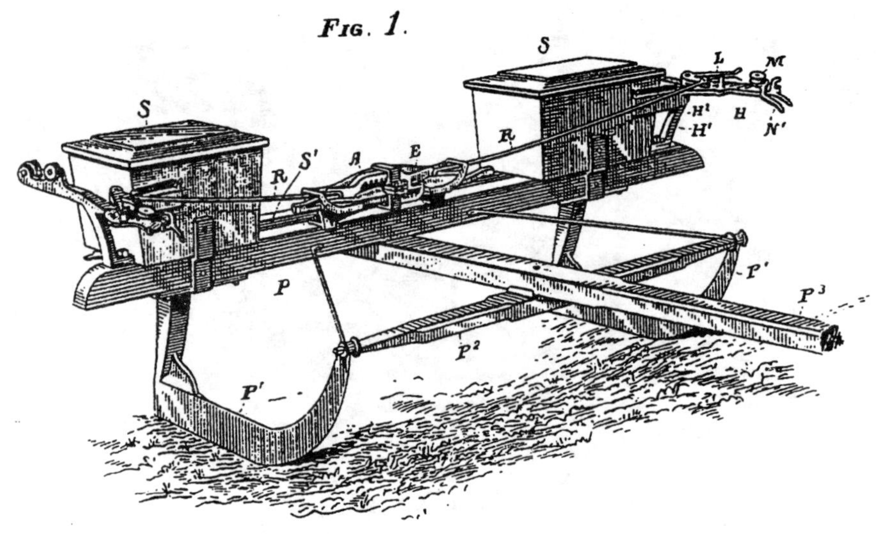

Peoria, Illinois

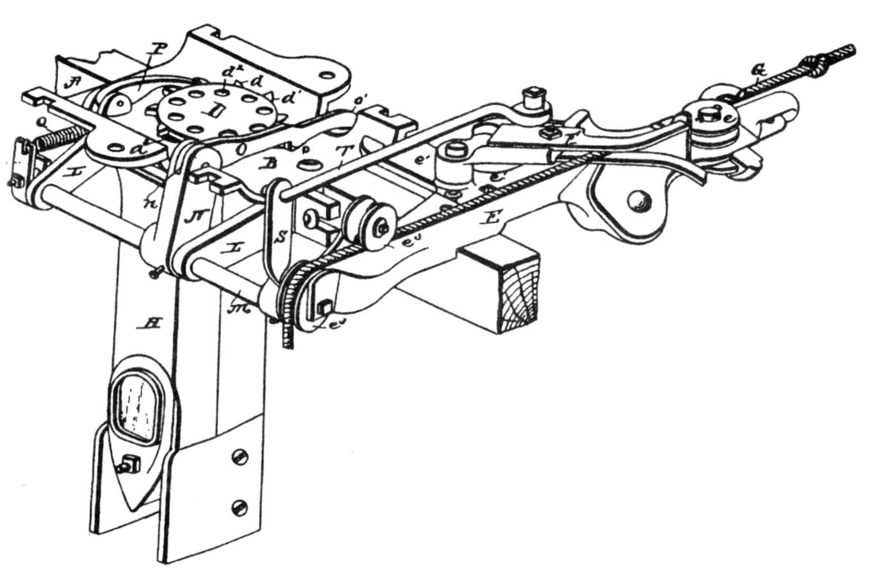

Madison, Wisconsin

October 22, 1889   C-R Planter/Marker   413,406

Aurelia, Iowa

December 17, 1889   Corn Planter   417,489

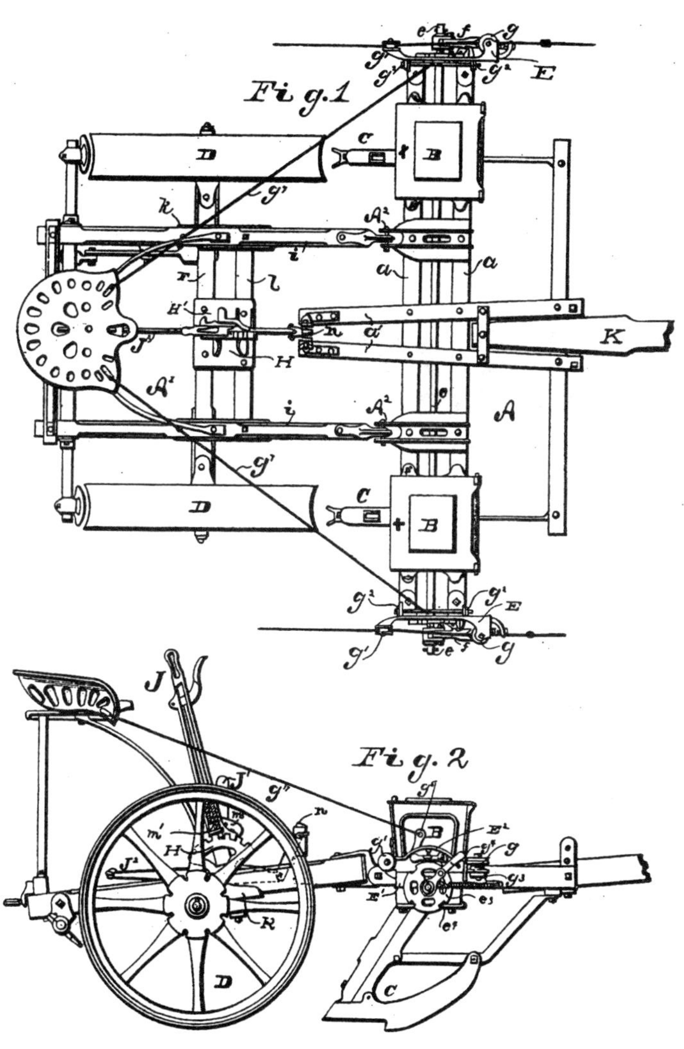

Troy, Ohio

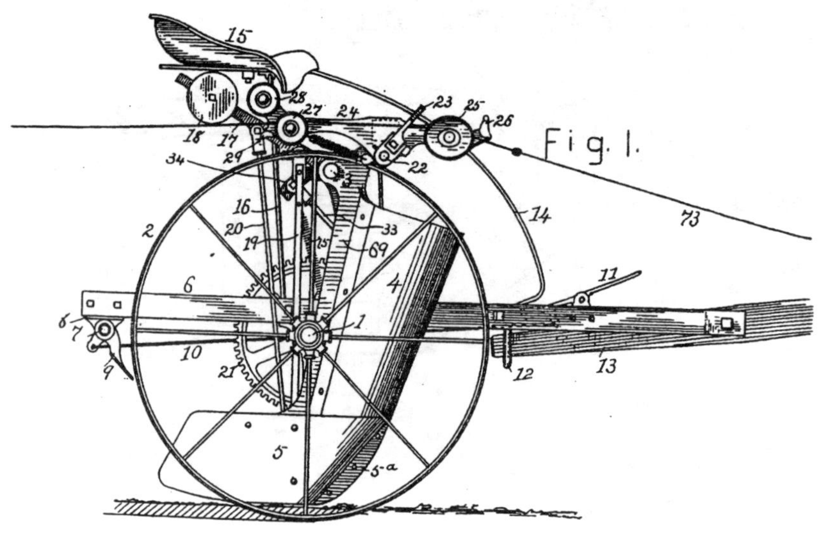

Decatur, Illinois

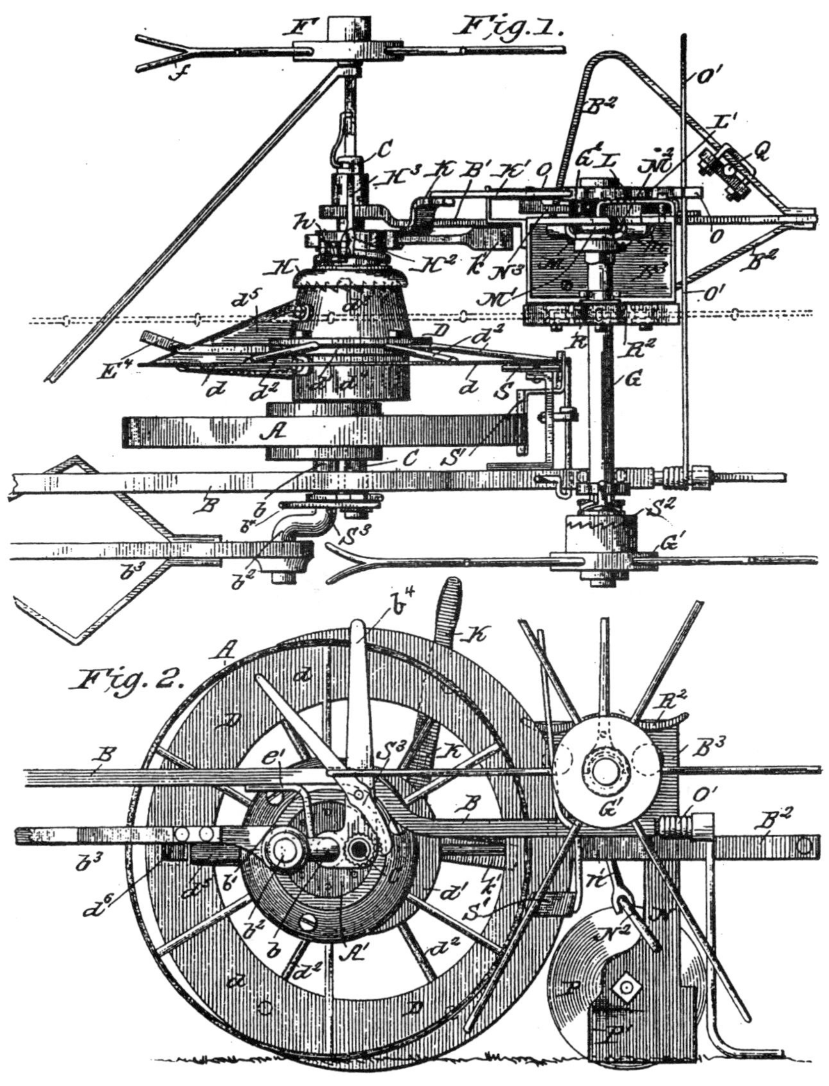

Moline, Illinois

July 3, 1894   C-R Corn Planter   522,381

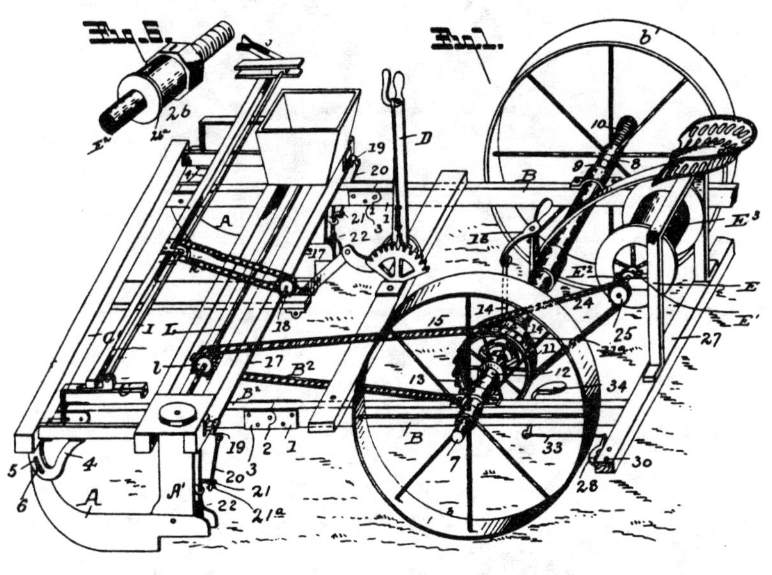

Cisco, Illinois

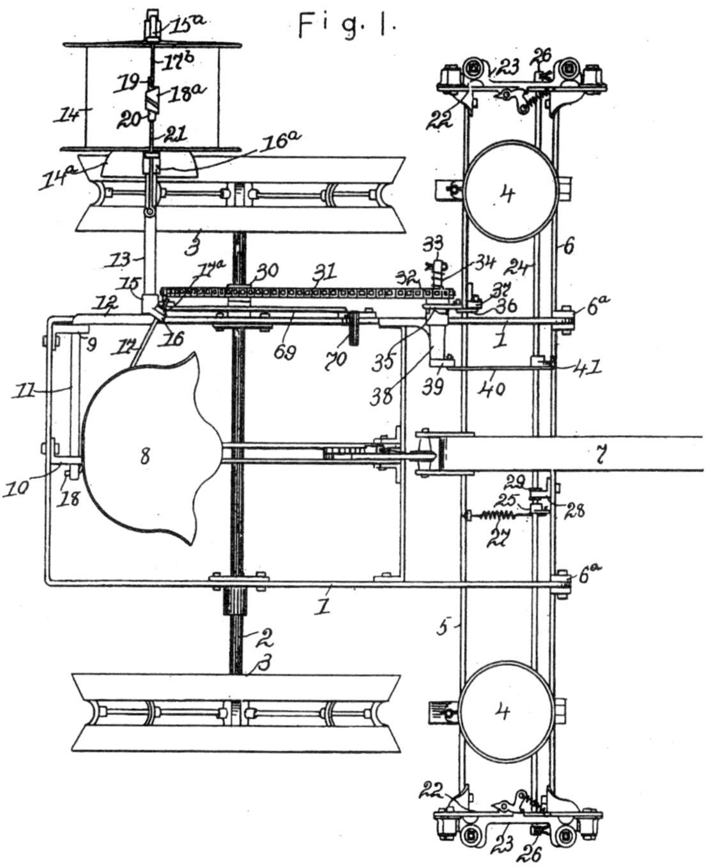

Canton, Illinois

ORSON D. PARK

May 8, 1900   Seed Planter   649,257

Manton, Michigan

September 4, 1900   Corn Planter   657,474

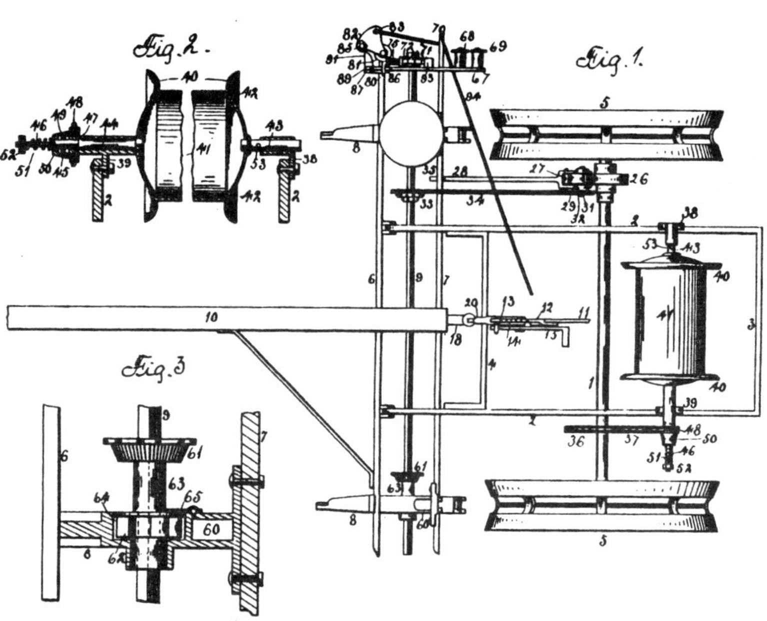

Moline, Illinois

October 16, 1900   Corn Planter   659,646

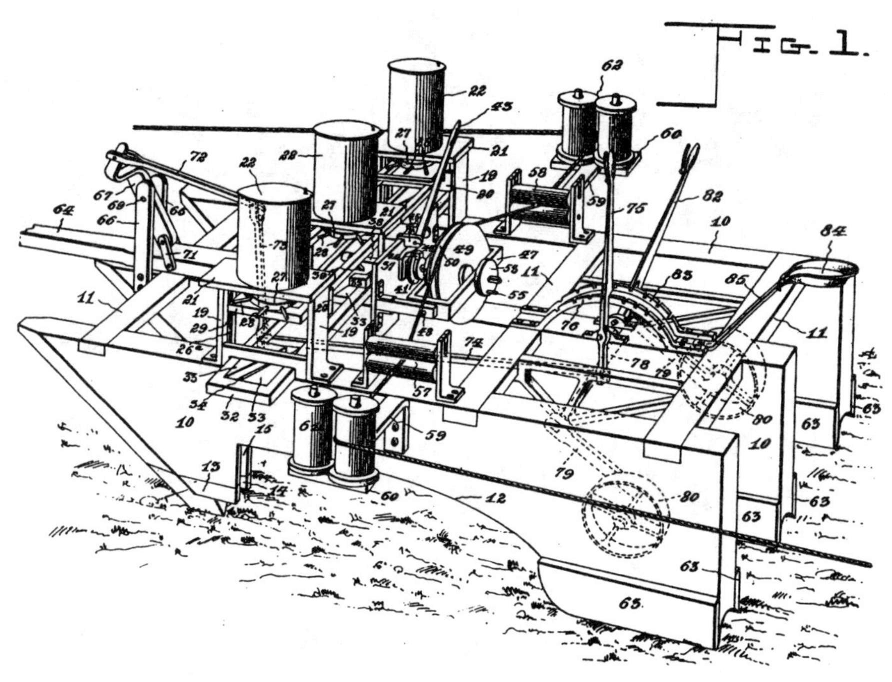

Newburg, Missouri

December 25, 1900   Corn Planter   664,767

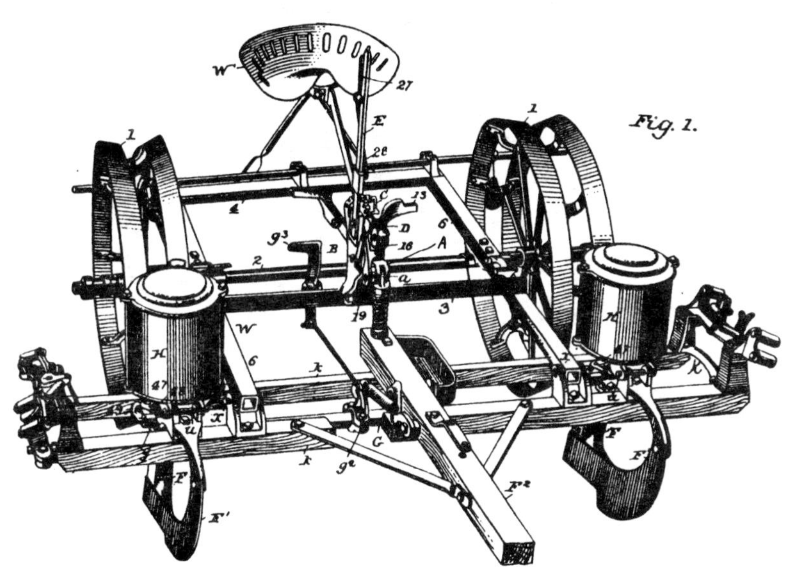

Rock Falls, Illinois

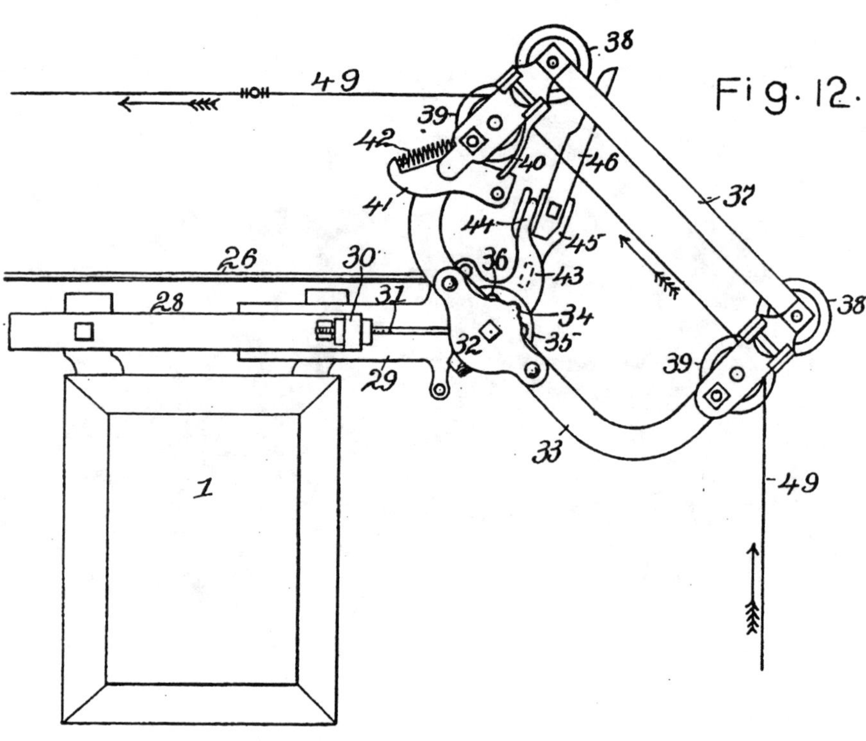

Chicago, Illinois

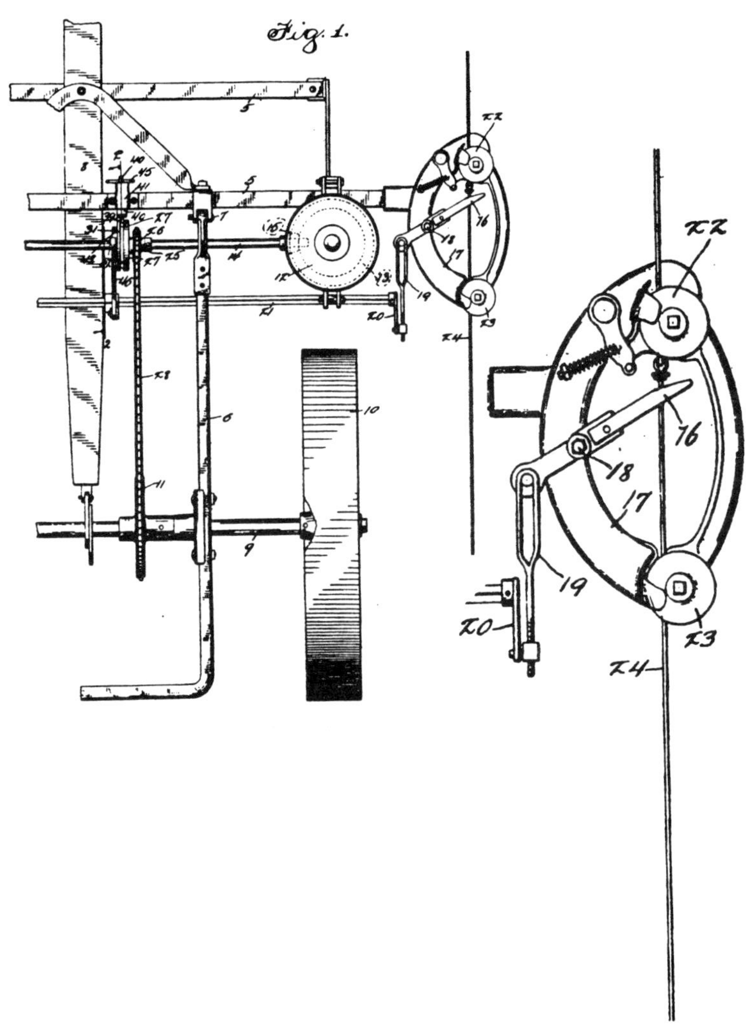

Fig. 1.

J.H. GROOTERS

June 14, 1904   Planter   762,706

Smooth wire cable drive seed drop

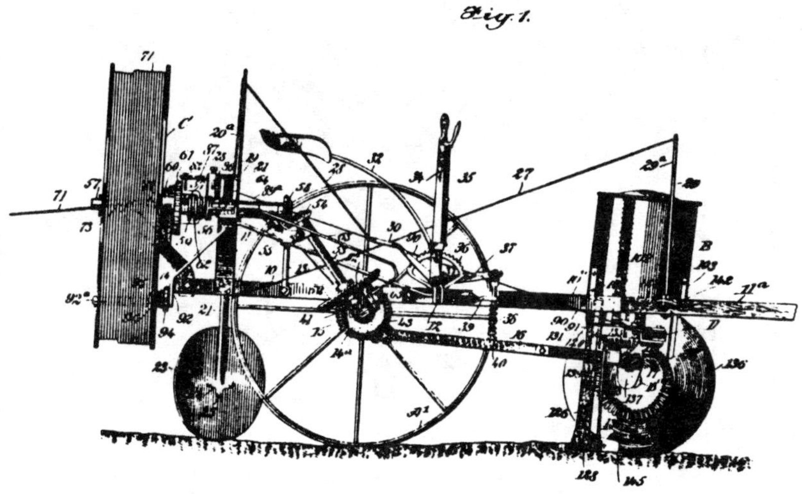

Allendorf, Iowa

August 23, 1904   Check-Row Planter   768,081

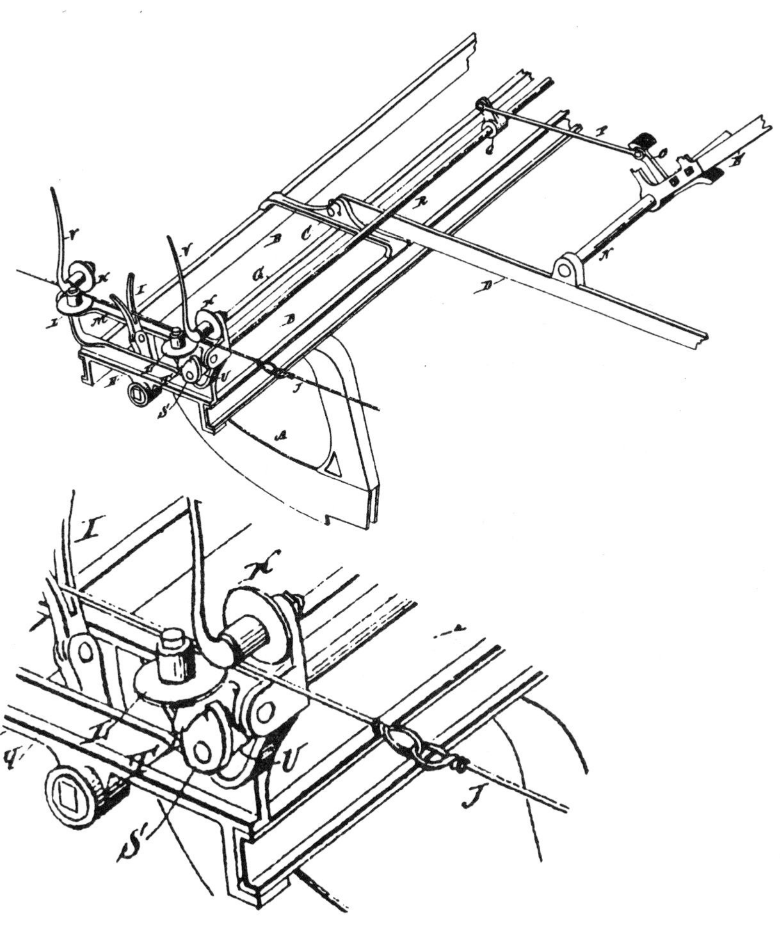

Clarence, Iowa

ALONZO M. CRISMAN

September 17, 1907   Check-Row Planter   866,339

A mechanical check-row planter and marker

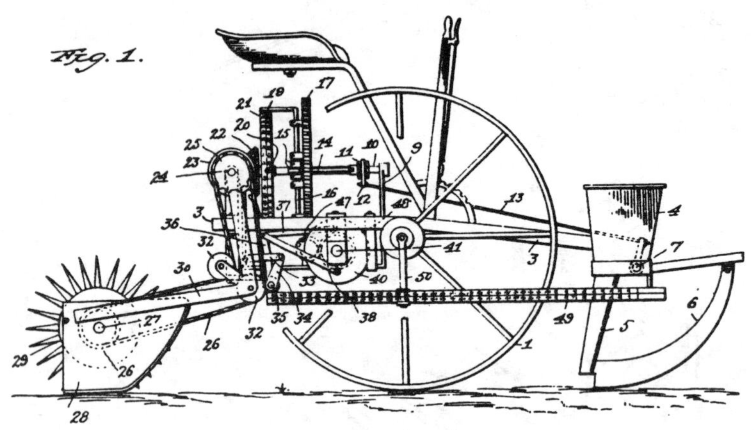

Rock Island, Illinois

CHRISTOPHER C. BUTCHER

January 7, 1908   Check-Row Planter   875,690

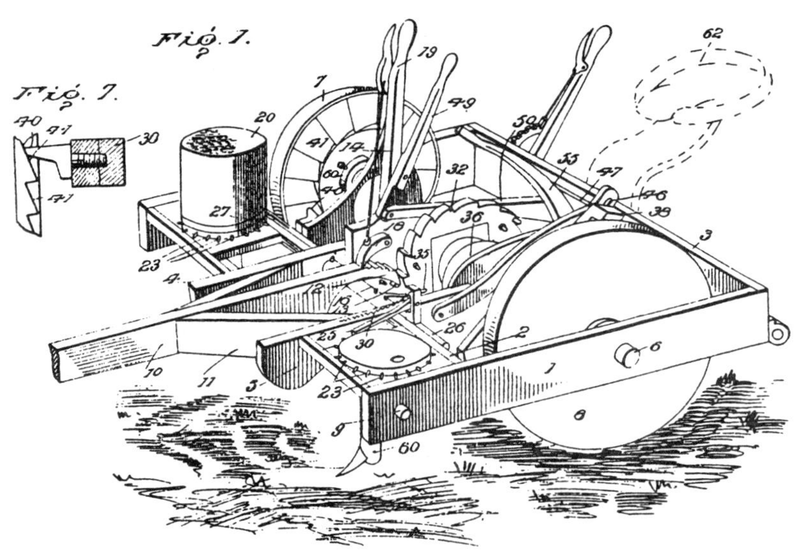

Urbana, Missouri

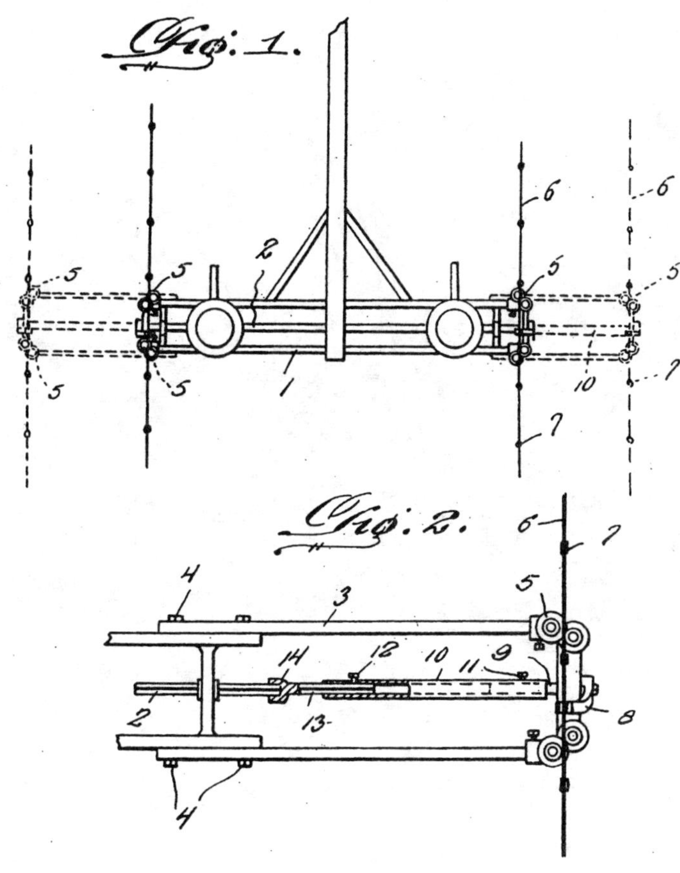

Milford, Nebraska

August 16, 1932   C-R Potato Planter   1,872,470

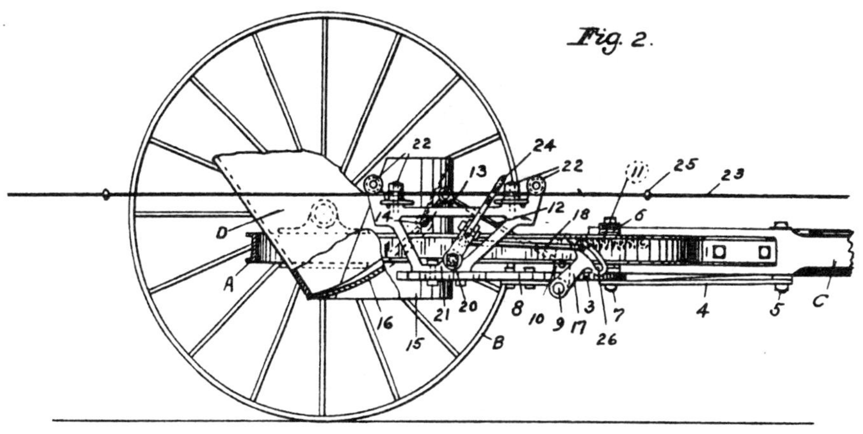

Shevlin, Minnesota

August 25, 1936   Check-Row Planter   2,052,455

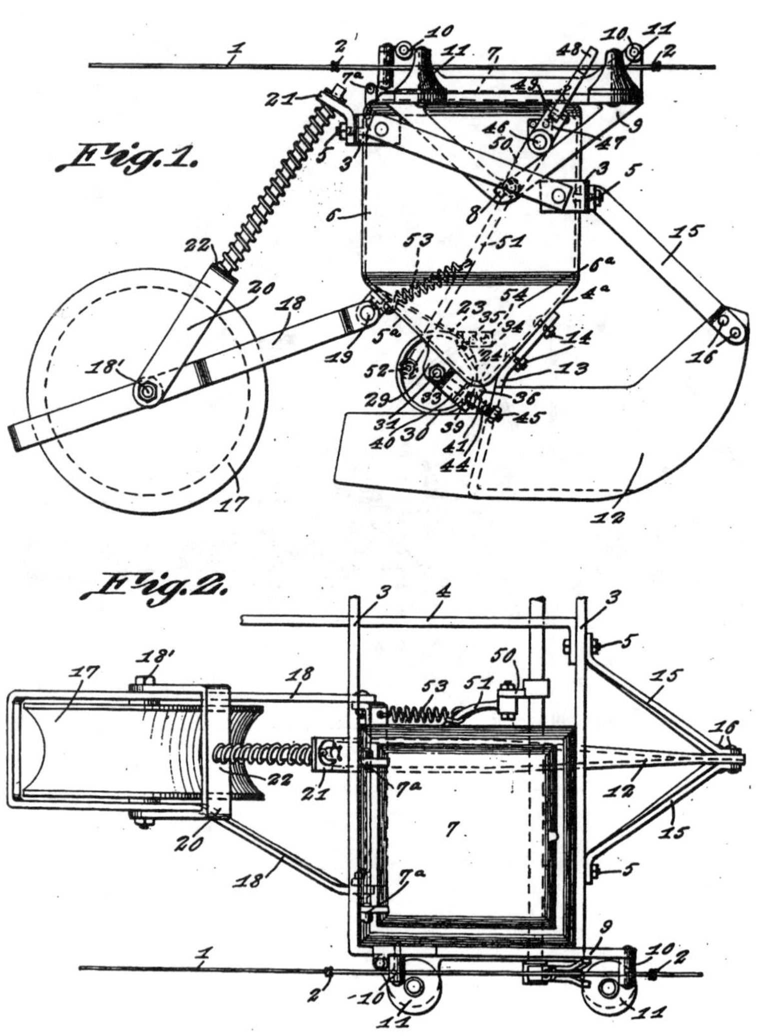

Clear Lake, Iowa

CLAUDE K. SHEDD AND EDGAR V. COLLINS

October 3, 1939   Check-Row Planter   2,175,035

A four-row planter; possibly experimental

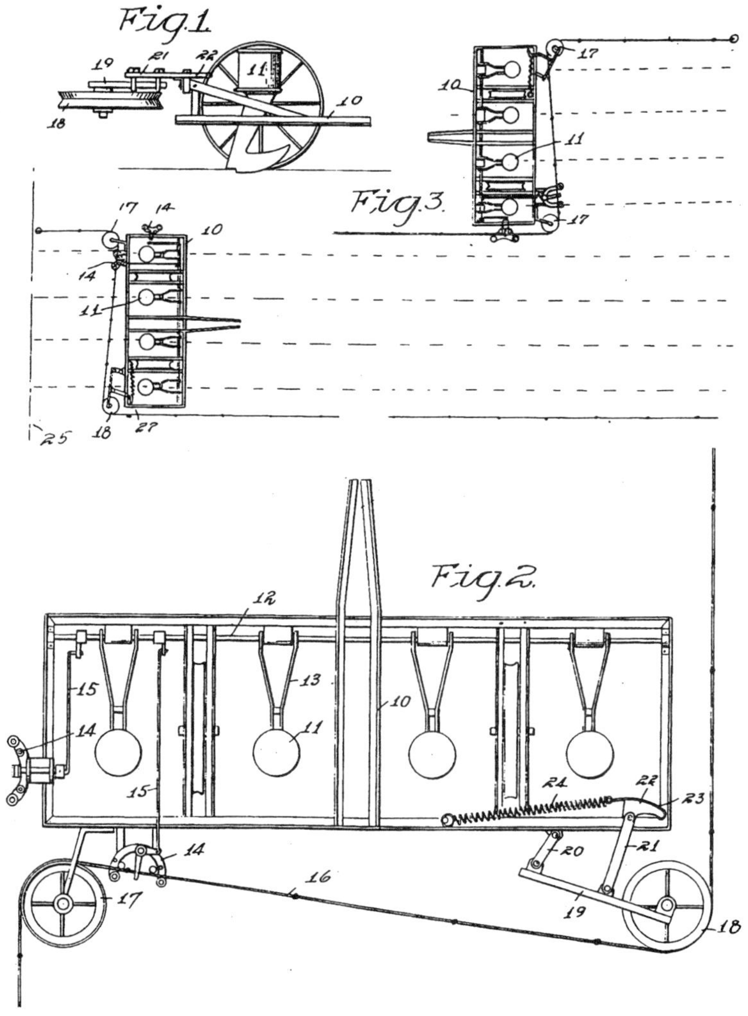

Ames, Iowa

SECTION 8

CHRONOLOGICAL INDEX
OF
CHECK-LINE AND KNOT PATENTS

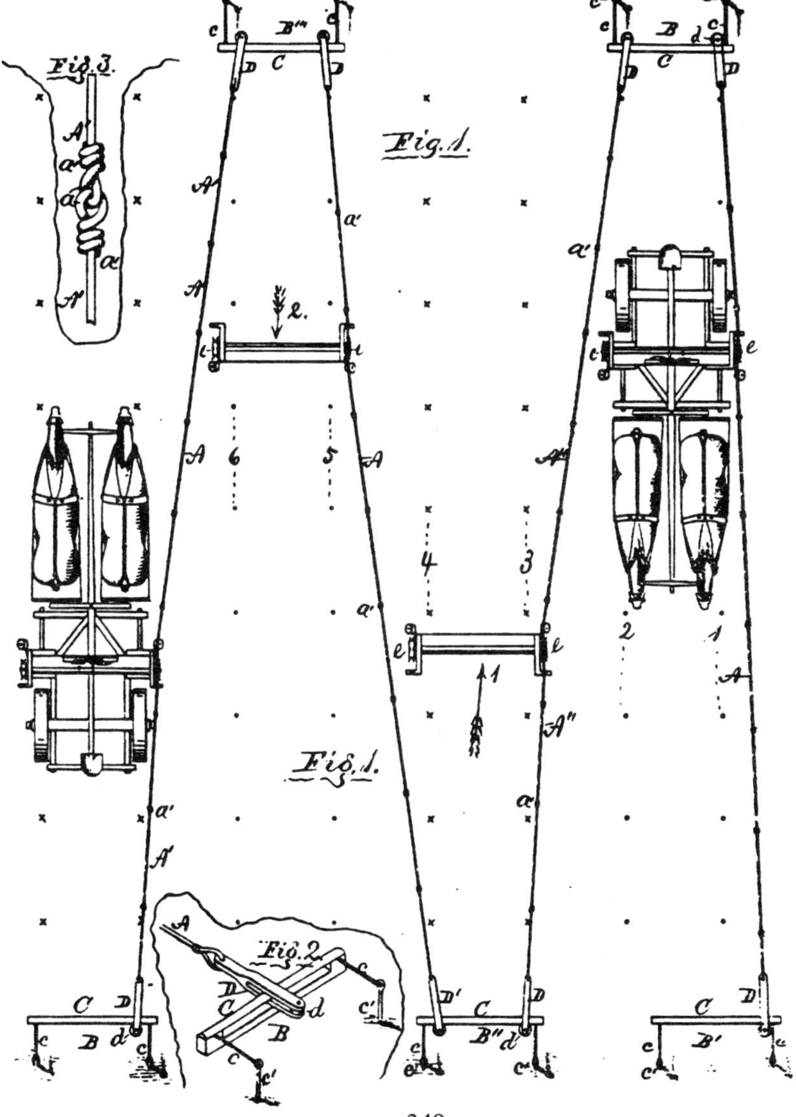

SECTION 9

INDEX
OF THE
CHECK-LINE INVENTORS

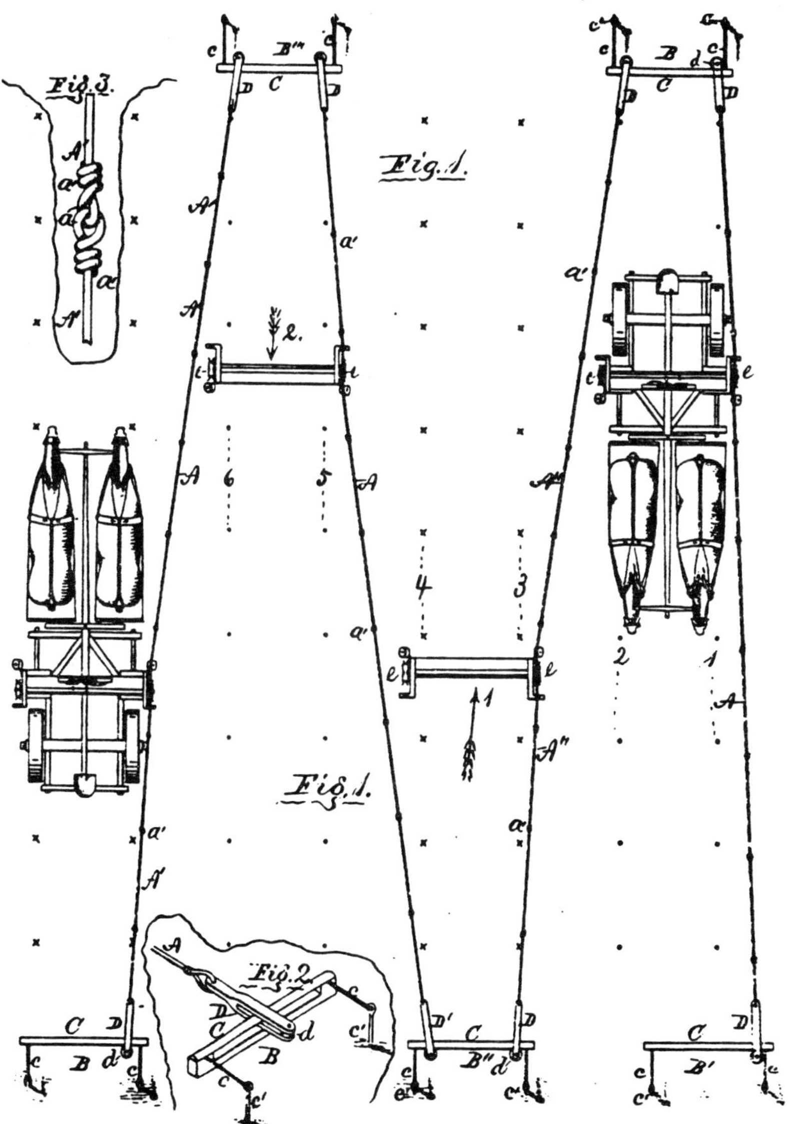

Index of the Check-Line Inventors

W

SECTION 10

INDEX
OF THE
ANCHORS AND REELS

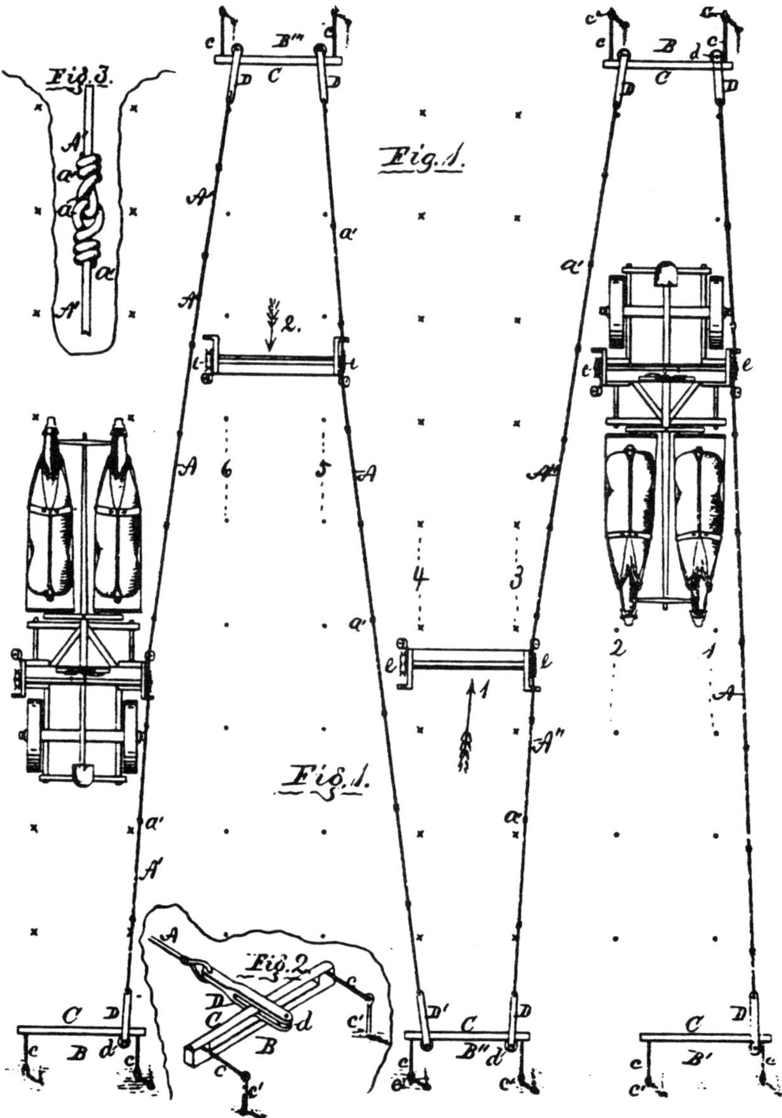

Index of Anchors and Reels

Index of Anchors and Reels

Index of Anchors and Reels

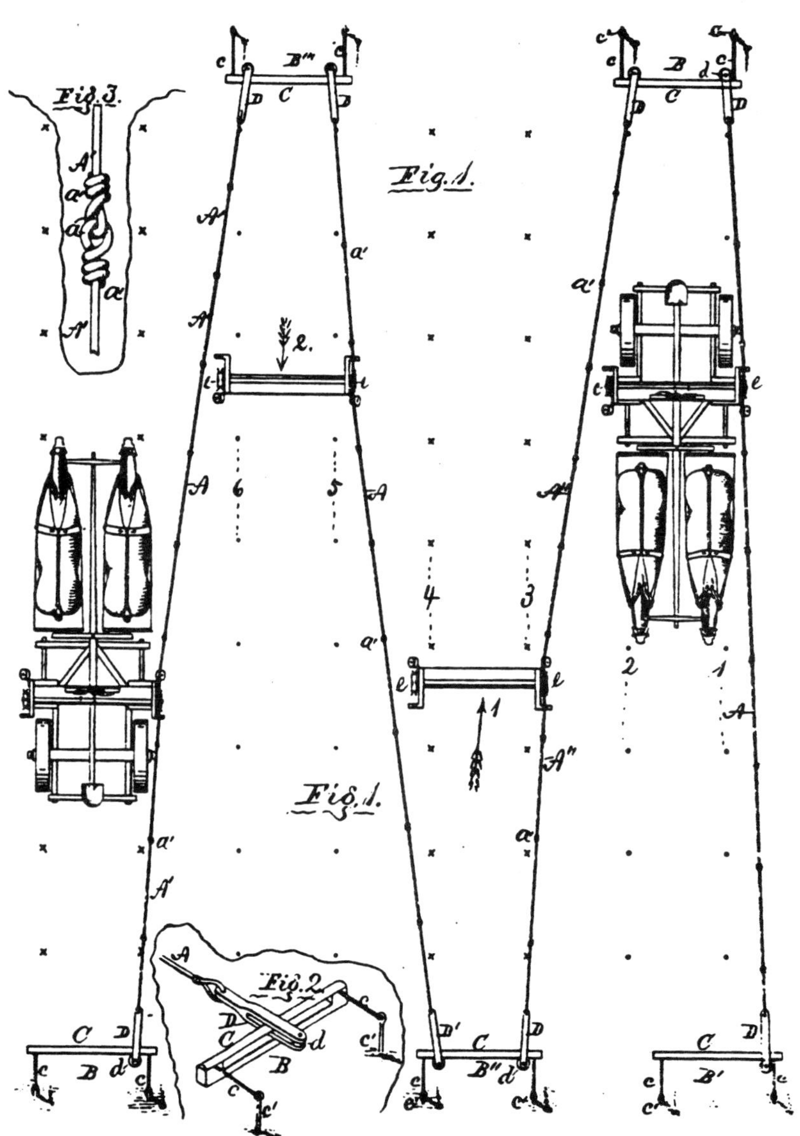
Fig.3.
Fig.1.
Fig.4.
Fig.2.

Sources

Information contained in this book came from the following sources:

U.S. Patent Office Report, 1850-1851

U.S. Patent Index, 1790-1872

U.S. Patent Gazette, 1872-1972

Selected U.S. Patents, 1842-1939

H.P. Deuscher Company, Farm Implement Booklet, 1891

J.I. Case Co. Inc., 32-G CR Planter Manual, 1929

Deere and Mansur Works, 999K Corn Planter Manual, 1935

International Harvester, McCormick CR-240 Manual, 1952

Wire Winder Mfg. Co., Spool Winder Ad

Hurt, R. Douglas, *American Farm Tools,* 1962

Hauff, William D., *Planter Wire*, 1970

Glover, Jack, *The "Bobbed" Wire Bible IX*, 1996

Planter Wire Collections of:
    Marvin Ginn, Morrowville, Kansas
    Al Greenwood, Kansas City, Kansas